Financial Policies and Management
of Agricultural Development Banks

Westview Replica Editions

The concept of Westview Replica Editions is a response to the continuing crisis in academic and informational publishing. Library budgets for books have been severely curtailed. Ever larger portions of general library budgets are being diverted from the purchase of books and used for data banks, computers, micromedia, and other methods of information retrieval. Interlibrary loan structures further reduce the edition sizes required to satisfy the needs of the scholarly community. Economic pressures on the university presses and the few private scholarly publishing companies have severely limited the capacity of the industry to properly serve the academic and research communities. As a result, many manuscripts dealing with important subjects, often representing the highest level of scholarship, are no longer economically viable publishing projects--or, if accepted for publication, are typically subject to lead times ranging from one to three years.

Westview Replica Editions are our practical solution to the problem. We accept a manuscript in camera-ready form, typed according to our specifications, and move it immediately into the production process. As always, the selection criteria include the importance of the subject, the work's contribution to scholarship, and its insight, originality of thought, and excellence of exposition. The responsibility for editing and proofreading lies with the author or sponsoring institution. We prepare chapter headings and display pages, file for copyright, and obtain Library of Congress Cataloging in Publication Data. A detailed manual contains simple instructions for preparing the final typescript, and our editorial staff is always available to answer questions.

The end result is a book printed on acid-free paper and bound in sturdy library-quality soft covers. We manufacture these books ourselves using equipment that does not require a lengthy make-ready process and that allows us to publish first editions of 300 to 600 copies and to reprint even smaller quantities as needed. Thus, we can produce Replica Editions quickly and can keep even very specialized books in print as long as there is a demand for them.

About the Book and Author

Financial Policies and Management
of Agricultural Development Banks
Carlos Pomareda

The effectiveness of many Agricultural Development Banks (ADBs) is seriously impaired because of their institutional design, specialization in lending at low interest rates, reliance on government subsidies and on low-cost funds from international financial agencies, high operating costs and low rate of loan recovery. To address these problems and to evaluate alternative policies and changes in strategy, Dr. Pomareda develops a multiperiod linear programming portfolio model and applies it to the case of the ADB of Panama. He concludes that the high cost of borrowed funds and the elimination of government subsidies could lead to the failure of ADBs unless they raise interest rates on loans, improve loan recovery, and perform multiple functions. Diversification in the sources of funds, with appropriate interest rate spreads may be the best path for ADBs to follow in today's financial environment and in the years ahead.

Carlos Pomareda is a research economist with the Inter-American Institute for Cooperation on Agriculture (IICA), San Jose, Costa Rica.

Dedicated to

My Parents, Jorge and Esther,
My Wife, Idalia,
My Children, Fabiola, Jorge, and Esther.

Financial Policies and Management of Agricultural Development Banks

Carlos Pomareda

Routledge
Taylor & Francis Group

LONDON AND NEW YORK

First published 1984 by Westview Press, Inc.

Published 2018 by Routledge
52 Vanderbilt Avenue, New York, NY 10017
2 Park Square, Milton Park, Abingdon, Oxon OX14 4RN

Routledge is an imprint of the Taylor & Francis Group, an informa business

Library of Congress Cataloging in Publication Data
Pomareda, Carlos.
    Financial policies and management of agricultural
development banks.
      (A Westview replica edition)
      Bibliography: p.
      1. Agricultural credit--Developing countries.
2. Development banks--Developing countries. I. Title.
HG2051 .D44P65   1984   332'.2   83-19674
ISBN 13: 978-0-367-01720-0 (hbk)
ISBN 13: 978-0-367-16707-3 (pbk)

Contents

Tables

Figures

Acknowledgments

This research was made possible with the encouragement, support and assistance of several people. I am thankful to Art Stoecker, Hong Lee, Sujit Roy, Robert Rouse, and David Upton for their valuable comments and suggestions, while I worked on my dissertation at Texas Tech University. I am grateful to José Emilio G. Araujo and José Alberto Torres for granting me leave from the Interamerican Institute for Cooperation on Agriculture (IICA), to complete my studies at Texas Tech; to Mike Gudger for his encouragement throughout my work; to Peter Hazell, Gustavo Arcia and Juan Antonio Aguirre for their comments; to Rocío Ramírez, Víctor Quiroga and Franklin Ureòa for their assistance in processing the data; and to Mayra Sequeira and Marta Sandino for typing the various drafts. Access to the data was guaranteed by the intervention of Virginia Velâsquez from the Agricultural Insurance Institute and the support of Generoso Pêrez from the Agricultural Development Bank of Panama.

I would have not been able to do this work wihout the constant patience and understanding of my wife and the faith and encouragement provided by my parents. To all of them I owe my achievements.

Carlos F. Pomareda Benel

Introduction

Credit for agriculture in developing countries is provided by commercial and development banks and by rural financial intermediaries, each serving a particular clientele. Commercial banks serve the largest farmers and operate under rigid commercial practices, characterized by, among other things, strong clientele selection criteria and market rates of interest. Development banks serve the small and medium commercial farmers and follow development oriented policies. Rural financial intermediaries serve the most isolated farmers and those that do not qualify as viable according to the criteria of formal lenders. The concern of this book is with development finance institutions and particularly with the Agricultural Development Banks (ADBs).

Because ADBs are an instrument of government policy for agricultural development, they have peculiar characteristics in their institutional design and they operate under subsidized interest rate policies. This study focusses in the nature of the institutional design and policies. They affect the banks' financial performance in the following way:

a. ADBs are limited to the issuance of agricultural loans while, on the liability side, they rely mostly on contributions from the governments and foreign loans under soft conditions. As a result they have a limited capacity to act as financial intermediaries.

b. ADBs operate under interest rates on issued loans that are below market rates. This limits their earnings and hence the rate of growth of loanable funds. These policies have been advocated for development purposes, but also they have been questioned because of the distortions they introduce in the capital markets.

c. As a general rule ADBs are characterized by low loan repayment. This low loan repayment is due in part to the instability of farmers' income because

1

of agricultural risks. But it is also due, to a great extent, to moral risks, and the inability of the banks to enforce their loan collection procedures.

d. ADBs issue a very large number of loans on a crop and site basis and therefore, face high operating costs.

e. Because of their institutional design, high costs and limited earnings on loans, ADBs have very slim resources. This low availability of resources and the large number of farmers that ADBs must serve, means that credit cannot be closely supervised. This contributes to the low loan recovery.

The above defines the generalized operating conditions of specialized ADBs. These conditions are to a great extent responsible for the limited growth of ADBs at the expense of their own resources, and hence the need for continuous government support.

If agricultural development is to be accelerated, one can expect this to be at the expense of a larger inflow of capital. If such are the needs, ADBs are expected to play a more meaningful role in the supply of financial resources for agriculture. The access to international soft loans and domestic subsidies is now more limited because of the countries' economic crisis and severe indebtedness. A change in strategy is needed for ADBs to generate their own funds and to improve their efficiency, and therefore; grow at a faster rate.

The specific problem with which this study was concerned was the evaluation of alternative financial policies and managerial decisions that would accelerate the ADBs' supply of credit. The interest is in how the growth of credit supply can be affected by: i) the bank attitudes towards risk and the enforcement of credit insurance, ii) the elimination of the requirement of serving small farmers, iii) the higher cost of funds and the elimination of government subsidies, iv) increased interest rates on loans, and v) the alternative of diversified banking

Given the situation described above, the objectives of this research were:

a. First, to develop an analytical framework that explains the performance of agricultural development banks in light of development purposes, institutional design and operating practices. This performance was also explained in relation to the risk in agricultural production and other facts that determine low loan repayment and hence contribute to reduced bank earnings.

b. The second objective was an analysis of changes in the prevailing conditions and how they affect the banks' growth. Various hypotheses are presented and tested for the case of a specialized ADB in Panama.

The achievement of the proposed objectives demanded two stages in the research and hence two methods of analysis:

a. In order to build an analytical framework for the operation of ADBs it was necessary to study their financial structure; their resource endowments, and the policies under which they operate, along with the nature of risks in agricultural production and other factors that affect farmers' loan repayment. To build this framework it was useful to examine the asset and liability structure of ADBs in Latin America, and to review previous studies.

b. The analysis of changes in current operating conditions and financial policies was achieved with the help of a multiperiod mathematical programming portfolio model for the ADB of Panama. Previous to designing and building the model, a considerable effort was devoted to a review of earlier models of bank portfolios. Since these earlier models were for commercial banks, the proposed model was enriched considerably through the incorporation of institutional and political constraints and measures of risk in the loan portfolio.

The model was used to simulate feasible changes in policy in the context on the Panamanian economic and agricultural sectors, and the organization of the Agricultural Development Bank of Panama (ADBP), and the Agricultural Insurance Institute (ISA). These changes related to the following issues:

a. As a public institution, the bank could well be interpreted as risk neutral, in the sense that any financial disaster could be overcome with government subsidies or additional soft funds. Increased risk aversion in the management of the bank would reflect a more careful administration of the funds.

b. The insurance program administered by ISA currently covers 30 percent of the bank's portfolio, however; if not supported by the government, ISA will not grow at a faster rate or it may even disappear. The benefits of credit insurance on the bank portfolio were analyzed in terms of loan recovery and administration costs.

c. The issuance of small loans on an individual basis increase the costs of administering bank credit. The policy of not providing credit to small farmers was analyzed in terms of the net effect on total costs and stability of bank earnings.

d. A large proportion of the bank's borrowings come from commercial banks, but with a government subsidy. An analysis was made of higher cost of borrowed funds, increasing the interest rate with an without government subsidies.

e. Diversified banking is feasible for the bank in the short run. It was analyzed by allowing the

issuance of bonds of different maturity and savings
and checking accounts of different sizes. The analysis also provided information on interest rate
sensitivity of borrowing, lending and purchase of
securities.

1
Agricultural Development Banking and the Supply of Credit: A Conceptual Framework

INTRODUCTION

The debate on the financing of agriculture in developing countries has long centered around two basic issues. One issue involves the organization and quality of service provided by the development finance institutions. The other issue involves the financial policies themselves.

Institutional design has made most agricultural development banks (ADBs) a class of rather specialized farm credit agencies. As such, they provide loans for agricultural production and few other financial services. Also, they rely mostly on international soft loans and government contributions, all of which contributes to their limited capacity to act as financial intermediaries.

Financial policies toward agriculture depart from the basic philosophy that low interest rates are a necessary condition for technical substitution and increased income in rural areas. There is, however, much controversy on the validity of these policies. They are, in part, responsible for a series of distortions in the capital markets, and the inability of the development banks to grow by generating their own resources.

ADBs have reduced earnings, because of interest rate policies and limited financial intermediation capacity. This is one of the major reasons for ADBs to provide a low quality service. But also, in order to fulfill development goals, ADBs must serve a large number of small farmers. This implies high operating costs for the banks.

If agricultural development is to be accelerated, one can anticipate an increased demand for capital. If such is the case, it is necessary to reappraise the financial policies toward agriculture and the role that agricultural development finance institutions must play. This appraisal is already under

way (World Bank/FDI, 1981), and new strategies are
being suggested.

Considering that the instability of farmers income due to crop failures is one reason for low loan
repayment, credit insurance is being considered among
the components of a new strategy to increase the supply of credit. Even though insurance could increase
loan recovery, the question still remains as to how
does it affect the bank's growth of credit supply in
comparison with alternative changes in policies and
management.

The remaining sections of this chapter analyze
the points outlined above, i.e., the institutional
design of ADBs and their financial policies; the
risks in agriculture and their effect on income
stability and loan repayment; the management of the
loan portfolio, the problem of loan recovery, and the
effects of credit insurance on the later. Ultimately, this chapter discusses basic concepts in the
management of an ADB, and the determinants of the
supply of credit.

AGRICULTURAL DEVELOPMENT BANKING: AN OVERVIEW

Development Purposes and Political Constraints

Development banks are the financial institutions
which integrate part of the system necessary to support economic development. As such, they have particular ways of fulfilling their functions, and also
if publicly owned, they are highly exposed to government intervention

Development banking emerged in the post-World War
II period to meet a need to supply low price capital
for economic growth. These banks are intended to
provide a complete package of services, including
capital and management for development purposes (Basu,
1974). Most development banks were created with the
purpose of serving a particular sector (industry or
agriculture), and hence specific types of development
projects. The latter are supposed to have high
social rates or return, but they also need low cost
capital to be financially viable.

Given this characteristic of development projects, banks face a conflict of purposes. Kane (1975)
explains that the conflict emerges because, as a
development institution, the bank should deal with
those projects with the highest ranking on the
development impact scale. As a banking institution,
it should finance those projects with the highest
ranking in the financial (interest rate) scale.

In deciding which development projects to finance, the development banks are influenced by

government goals and policies, and by financial criteria. Governments exercise pressure on the banks to finance particular projects expected to benefit target groups. However, the lower the expected monetary return of the project, the more difficult it is to get the funds to finance it. To the extent that the government wishes to reach certain political targets and groups, it will increase the level of subsidy, and/or the pressure to get external-low cost funds. While fulfilling these functions, a development bank becomes a mere conduit for funds and less of a financial intermediary.

The above functioning of development banks has been criticized. The criticism is more severe on public development banks than on private or mixed capital development banks[1]. To the extent that the bank leans more towards private ownership, profit and hence monetary return on projects becomes a more important criteria in project financing. In this regard, Kane (1975) concludes that development banks, therefore; make a more significant contribution to economic development than private banks. This assertion is questionable in the long run, when banks with low earnings have a slower growth as a function of their financial performance, and hence the need for continuous subsidies.

This discussion is indicative that development banks intend to operate as banks within the limits imposed by political constraints and institutional design. They are concerned about earnings while fulfilling development goals; hence at the difference of commercial banks, profit per se is not the motive in development banking. Recognition of this environment is important when making an evaluation of performance of development banks.

Institutional Design of Agricultural Development Banks

In a political context, ADBs are an instrument of government policy for agriculture. As such they should serve particular groups of producers, usually the small and medium commercial farmers, while excluding the smallest subsistence farmers. They supply credit for crops that have high priority, either as part of food supply programs or for those that provide the basic foreign exchange earnings. They are characterized by very large operating costs, because of the type of clientele they serve. Finally, as a general rule, they have poor loan collection performance, which reduces even further the earnings margin or makes it negative, hence the permanent need for government subsidies[2].

There has been a strong belief that ADBs as well as other development banks, should be specialized institutions. In fact, it is possible and rational for an ADB to specialize in lending to the agricultural sector; but that is not to say that the bank should specialize in being a lending agency and not playing the role of a financial intermediary. This misconception has led to the design of institutions with a very peculiar structure. In fact, Von Pischke, Hefferman, and Adams (1981) refer to them as "specialized farm credit institutions." The great majority of them are publicly owned banks, limited to offering farmers low interest rate loans, but no other financial services. They do not accept checking and savings deposits, provide money transfer services, store valuables for safe-keeping or serve as fiduciaries. Therefore, their only sources of funds are loan recovery, domestic borrowings from the Central Bank, and external borrowings from International Financial Agencies, usually at very low rates and long deferment and repayment periods.

The limited capacity to access market funds results in alienation of the institution, because it can not intermediate between rural savers and borrowers, and it limits itself to serve as a link between the government and the rural sector (Von Pischke, 1981). On the other hand, this institutional design and the high operating costs do not allow the bank to offer good quality credit, hence the farmers' preference for rural private lenders (Ladman, 1981).

A significant portion of the financial resources for agriculture is provided, however, by non-specialized (public, private and mixed ownership) development banks. Because of the structure of their asset and liability portfolios, these institutions have better possibilities for acting as financial intermediaries. On the liability side they look much like commercial banks, since they borrow from internal and external sources, and receive demand and time deposits. On the asset side they serve various sectors (althoug they may concentrate on agriculture) and they invest in securities and issue loans of different maturity and risk, hence allowing for more flexibility in the management of the portfolio.

In a recent analysis of the portfolio composition of 97 development banks in Latin America, Pomareda (1982.b) found very peculiar characteristics of those banks serving primarily or exclusively the agricultural sector (see Table 1.1). Banks with over 90 percent of their resources allocated to agriculture were exclusively public banks; they were smaller than the other banks and they depend fundamentally upon internal resources. The most significant contribution to the latter was public borrowings. The proportion of public deposits in their portfolio was around 3 percent compared with 40 percent for other banks.

The discussion presented here suggests that the ADBs have much to gain from acting as financial intertermediaries. This is to say that the banks can restructure the composition of their assets and liabilities, but still specialize in lending to agriculture[4]. A bank could even charge low rates on certain agricultural loans, if it can earn more in its role as financial intermediary by issuing checking and savings accounts and investing on securities.

Financial Policies Towards Agriculture

As part of the same philosophy of finance for development, interest rates for agriculture are below market rates. Most developing countries provide subsidized interest rates to agriculture, with the main purpose of inducing the adoption of capital intensive technologies that would result in increased productivity. Low interest rates have been visualized as a necessary condition for agricultural development, yet much controversy exists on the subject. Besides the criterion of "low interest rates to induce technology adoption," several other arguments are offered to justify this policy. Some of these arguments are discussed below.

Low interest credit is offered as an alternative to high cost funds supplied by informal lenders in the rural markets. These groups are believed to exercise monopoly power, and hence, to receive returns above their costs. Nevertheless, informal lenders usually offer to farmers other services like input supply and a guarantee of purchasing the harvest (Barton, 1977; Bouman, 1979), hence justifying a higher cost of capital. On the other hand, the high cost of informal credit, usually delivered at the farm, may not be higher than the real cost of official credit which includes the farmer's time until the credit is obtained and during the loan supervision period (Adams, 1981). In many cases, however, these intermediaries do exploit the opportunities in the rural sector and exercise monopoly power, particularly among the less fortunate farmers who do not qualify as credit worthy according to the ADB criteria.[5]

Perhaps the strongest argument for low interest credit has its roots in the historical time when development policies were originally designed. The development philosophy gained strenth in the 30's when the world recession implied negative real rates of interest. Therefore, it did not seem strange to

Table 1.1
Average Financial Structure of Development Banks in Latin America, 1975-1980

Variable	Percentage of Resources Allocated to Agriculture			
	0-10	10-50	50-90	90-100
Number of banks	53	43	13	9
Percentage of public ownership	58.26	55.60	75.54	100
Total assets (Million US$)	826.94	1,239.356	3,159.136	92.56
Structure of Resources (%)	100.00	100.00	93.00	99.56
Capital	24.83	19.69	29.74	48.89
Internal sources	62.70	65.37	59.89	42.07
External sources	12.73	15.67	8.76	8.59
Distribution of Internal Sources (%)	100.00	100.00	100.00	100.00
Bonds and values	10.22	11.28	2.60	2.49
Public deposits	35.69	36.59	47.40	3.14
Public borrowing	34.78	35.46	32.25	52.46
Private borrowing	5.91	5.84	4.77	15.13
Other	13.40	10.83	12.98	26.78
Allocations by Sector (%)	100.00	100.00	100.00	100.00
Agriculture	2.47	27.65	66.82	98.13
Industry	38.85	34.66	14.05	0.00
Commerce	5.22	3.67	1.98	0.00
Construction	23.98	8.41	3.13	0.31
Other	29.48	25.61	14.02	1.56
Total contribution of all banks to agriculture (Million US$)	884.41	4,856.76	11,502.53	658.98

Source: Pomareda, C., 1982.b .

Note: It excludes the Banco do Nordeste, Brazil, the largest development bank in Latin America.

offer development finance at 2 or 3 percent interest
rates. However, if we focus on real rates of interest, the rate on the loan should include at least the
cost of inflation, hence higher rates. Failure to do
so will result in decapitalization of the banks.
Such has been the case for many Latin American countries which failed to reconcile their financial
policies and current inflation rates (Galbis, 1981).
However to the extent that there is money illusion
and segmented capital markets, savings are movilized
in Latin America even with negative interest rates.
This has been an argument against using real interest
rates on loans.

When inflation was not so severe and when international financial agencies had a stronger position,
they could lend at very low rates. It was believed
therefore, that domestic development banks should
provide farm credit at the same rates. That, however, ignores the administrative costs of credit
because of the rather large number of small loans.
If ADBs act as banks, they may have the right to
transfer those costs to the borrowers; on the other
hand, if they act as instruments of government policy, then they can expect government subsidies.

The higher the cost of capital, the lower the expected profitability of the financed enterprise, and
hence, a smaller margin to the farmer. Profit is
therefore, a determinant of loan bearing capacity,
and it is believed that lower interest rates increase
profit margins, and hence loan repayment ability. Low
rates, however, induce misuse of credit obtained for
agriculture but invested in alternative projects. As
a result, a farmer could have an excellent record on
loan repayment because of higher returns to the
borrowed money put in other uses, but not because of
a larger profit margin in agriculture.

One of the strongest and most debated arguments
for low interest credit to agriculture is the income
distribution effects, expected to benefit the rural
poor. This, as recognized by Gonzalez-Vega (1977 and
1981), assumes the larger benefits to be distributed
among a large number of small producers. In practice, however, even though ADBs show a large number
of loans, the number of beneficiaries is much
smaller. The reason for this is that loans are provided on a crop-site basis. Hence, a large commercial farmer, with several properties and growing
various crops, may receive five or more of the
largest loans, while small farmers receive one or at
the most two small loans.

Much debate still exists on these issues. Many
governments in Latin America are in the process of
revising their interest rate policies for agriculture
in order to keep institutions financially viable, and

able to manage in the current inflationary process (ALIDE, 1981). Nevertheless, there are some countries willing to continue subsidizing agricultural credit for political reasons and fear that higher rates may result in a decline in credit demand and lower productivity in agriculture. Low interest rates to agriculture are being claimed nowadays to compensate farmers for an unfair externally influenced rise in input prices and low domestic and export product prices, which reduce farmers income.

An issue of relevance in agricultural finance is therefore, the responsiveness of farmers to higher interest rates. It is argued that the elasticity of demand for public credit is rather insensitive to changes of the nominal interest rate, because the later is only a small portion of the total cost of credit the farmer faces (Adams, 1981). Furthermore, this sensitivity could decrease if better quality loan services are provided and larger volumes of credit made available.

An interesting paradox exists on interest rate policies and agricultural risks. Low interest rates to agriculture have been justified from the farmers' point of view, because of high risk-low profit of agricultural enterprises. Low interest is therefore, expected to compensate for the cost of risk. However, from the bank's point of view, as a financial institution, it should charge a higher interest rate on loans to the riskier enterprises, i.e., a higher rate for agricultural loans. This paradox and the naturally expected high default on agricultural development bank loans, is an important reason for the ADBs limited growth when they depend on their own resources.

In the case of countries where inflation has been low for a long period of time but suddenly becomes a major drawback on the economy, its psychological effects may be more pervasive on credit demand than anticipated. When interest rates rise rapidly they have a strong effect on decreasing desided investment, and hence, on the demand for credit. A case in point was recently observed in Costa Rica (January 1982). When official interest rates rose from 12 to 20 percent, to adjust partially for inflation, the demand for agricultural credit declined but only temporarily.

While subsidized interest rates may not be justified from a financial point of view, there are other reasons why at a particular point in time agricultural interest rates may need to be low. If that is the case, the banks should then be prepared to supervise agricultural credit, for farmers to use it in the desired investments and not outside agriculture. However, this increases the banks' operating costs. In this case, the ADBs should be prepared to generate financial resources from other activities, in order to allow themselves to fulfill their development goals.

RISKS IN AGRICULTURAL PRODUCTION AND LOAN REPAYMENT

The Nature of Risks in Agriculture

Agriculture is a risky enterprise and risk averse behavior among agricultural producers has been given as one explanation for low investment, limited technique adoption and the slower growth of agriculture compared with other sectors. The situation is more severe in developing countries, where other factors aggravate the effects of risks.

Risk in agriculture stems from various sources. First, uncertain input supplies and prices make production costs a random variable. Second, uncertain yields and product prices imply risk on gross returns. There are also risks because of storage and marketing loses that the farmer often has to sustain. In order to manage risks, farmers adopt different strategies, including crop and technology diversification, reluctance to use modern inputs and credit, and the use of agricultural insurance.

Uncertain input prices are not widely recognized in the literature as a primary source of risk. However, experience shows that to avoid crop damage from diseases or pests, farmers purchase insecticides and fungicides. Under unexpected widespread phenomena, the price of these inputs will rise to very high levels because of short term inelasticity of supply. The use of insecticides and pesticides can reduce risks; however, Just and Pope (1979) have demonstrated that a risk averse farmer will tend to over-invest in such inputs and this can be just as socially inefficient as under-investing in inputs which increase risks (as fertilizers). There is also often the case that in spite of the farmers' willingness to use certain inputs as part of a modern technology, these can not be obtained. In fact the success of some rural change and development projects has been guaranteed thanks to the provision in kind of those inputs. Such situations have been reported by Scobie and Franklin (1977), and by Ccama and Pastor (1982).

Yield variability is a common source of risk and it is as significant in the arid environments as it is in the humid and subhumid tropics. It is usually associated with hail, frost, drought, fire, dust storms, hurricanes and river floods. Also, diseases and inappropriate use of technologies can result in

loss. Yield variability associated with climatic factors is widely documented in the agronomy and agricultural economics literature, as shown for example by
Anderson, Dillon, and Hardaker (1977). Yield variability associated with higher levels of input use is
also evidenced in the works of de Janvry (1972), and
Moscardi and de Janvry (1977). Yield variability is
an important reason for low rates of technology
adoption as explained by Berry (1977), Green (1978),
and Binswanger (1978). The risks of agricultural
production emerging out of yield variability have
provided the rationale for crop insurance; but this
will be demanded by farmers only to the extent that
at the price offered (premium) it is more cost effective than traditional methods of production risk
management.

Price risk has been given greater attention in
the literature, particularly in developed countries.
The U.S. agriculture price support programs are an
indication of the importance of the issue. The income stabilization effect of price support programs
in U.S. agriculture was examined by Baker and Dunn
(1979), and Gardner (1979); concluding that such programs affect positively the financial viability of
farms. In developing countries price support programs for cereals are widely used, yet they have been
strongly criticized because of their distortive
effects.

Much of the research on price variability has
been with regard to its effects on consumers welfare,
as reported in the works of Waugh (1974), Subotnik
and Houck (1976), Masell (1969), Just et. al (1977),
among others. On the producers side, the desirability of price stabilization has been demonstrated by Hazell and Scandizzo (1975), among others.
Agricultural price stabilization programs in national
and international schemes have been widely advocated.
However, few of these have operated effectively
because of their large costs and lack of political
feasibility (Hazell and Pomareda, 1981).

This analysis suggests that there are several
sources of risk in agricultural production and similarly several ad-hoc ways of handling them. Clearly,
agricultural insurance is only one way of contribut-
ing towards the stabilization of farmers' income when
yield failure occurs.

Income Stability and Loan Repayment

The allocation of financial and physical resources at the farm level can be examined with refer-
ences to Figure 2.1.[6] Following the principle of
money fungibility, the different money sources are

aggregated into a capital input, which is in turn assigned to production processes according to the farmer's decision criteria.[7] Farmers, especially those that are smaller in terms of income, ussually combine the earnings from several activities in one account and use those resources according to priorities. The priorities could change given the circumstances; hence money (earned and borrowed) could be used for present consumption and investment outside agriculture, both favored by the low interest rates at which official agricultural credit is obtained.[8]

As a part of the decision making and production process shown in Figure 2.1, money is used to purchase agricultural inputs and, as discussed before, here is the first origin of risk. In addition, it is important that inputs are used at the optimal time, because this affects the impact of weather factors on the variability of crop yields, which is the second source of risk. The last source of risk in the cycle is market risk, reflected in the instability of prices. It is clear that net income (at the far right of Figure 2.1 is a random variable. More over, beyond this point are additional elemens that determine the available funds to repay the bank loans. Farmers allocate net income into planned consumption, savings and payment of outstanding debts. It is common to find that farmers would engage in luxury consumption even before paying outstanding debts, or else that present consumption is given greater importance than future consumption (savings).

This process of the allocation of funds in a risky environment at each stage of the production process, explains why farmers may be unable or unwilling to repay their loans. The banks' awareness of this process for each individual borrower would provide the basis for loan provisions. However, the bank can not do much to improve the immediate position of those not qualifying for loans. The bank can however, request government action to supply inputs at the opportune time and to provide support to input and product prices or the provision of agricultural insurance.

This simplified analysis of income variability at the farm level provides the rationale for income stabilization policies of different kinds. However, a point worth emphasizing is that farm income can be stabilized and that should increase the farmers' debt repayment capacity. Yet, the removal of all income variability by itself does not guarantee loan recovery because possible moral risk, which is induced in part by the bank's weakness to inforce collateral requirements or to collect loans.

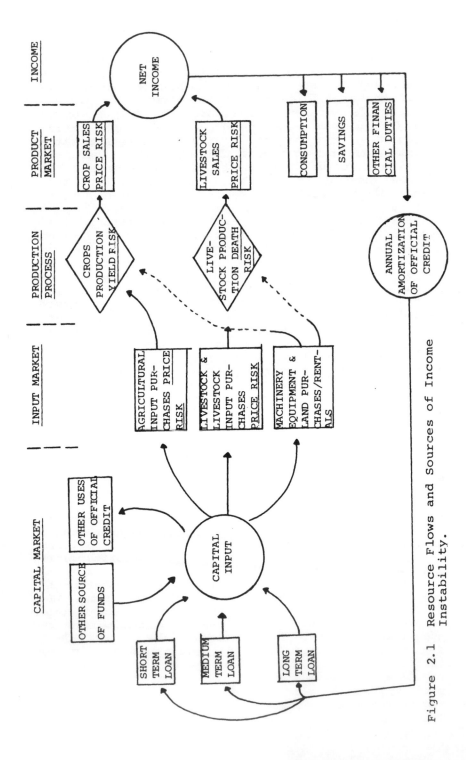

Figure 2.1 Resource Flows and Sources of Income Instability.

Agricultural Insurance and Credit Insurance

As discussed, farmers confront numerous risks throught the growing season. Within this array of risks, there are specific measures and policies to deal with each one. Agricultural insurance has been devised with the specific objective of compensating farmers against yield failure. By purchasing an insurance policy the farmer protects the value of his crop harvest or animal against specific disasters. In principle the coverage can be for as much as 100 percent of the value. Expected product prices are used to calculate the value of the output or the total coverage.

Agricultural insurance schemes are widely difused in the developed countries. The U.S. Federal Crop Insurance Program has more than forty years of experience and covers more than thirty crops against most natural disasters (hail, drought, flood, fire, and others). Long existing programs are also in Canada, Sweden, Israel, Japan, and Australia. In the developing world, agricultural insurance is rather new, except for Mexico and Puerto Rico where the programs exist since the 1950's. Of all the programs referred to above, only the case of Mexico is a credit insurance program

Credit insurance is slightly different than agricultural insurance. It protects only the value of the loan and not the total value of the harvest. By purchasing a credit insurance policy, the farmer protects his loan or a portion thereof. If the harvest (or part of it) is lost, or if the animal dies, the insurance agency pays the bank the amount due by the farmer, thus allowing him to return to production without seriously decapitalizing his resources.

On theoretical grounds, one can discuss the benefits of credit insurance from various points of view. Some of these benefits as well as the costs have only recently begun to be tested (Pomareda, 1981.a; Ccama and Pastor, 1982; IICA, 1981). Since credit insurance pays the farmer's debt in the event of yield failure, it can stabilize the farmer's income and increase his debt bearing capacity. By paying his bank debt, it allows the farmer to return to the bank in the following year and request a new loan and continue investing. Within the agricultural system, insurance spreads risks among farmers, regions and crops over time and it precludes the need to expensive ad-hoc disaster efforts, as the insurance system itself is able to offset losses from the reserves collected in good years and in unaffected areas.

Although highly favored as part of a rural development strategy in developing countries (Ray, 1974; Tewary and Sharma, 1978; Koropecky, 1980; and Gudger, 1980), crop insurance has been questioned on two grounds. First, its justification is questionable when the variability of yields is small and when traditional methods of risk management prove to be effective. Second, its feasibility is limited by its costs, particularly if premiums may need to be too high for the program to be financially self sufficient or else need to recur to government subsidies (Roumasset, 1979; Crawford, 1979). The later point has been a major reason for debate with regard to agricultural insurance and agricultural credit insurance. The concern is valid, particularly when considering that agricultural development banks (especially those servicing a large number of small farmers) engage in large administrative expenses. An insurance agency servicing the same clientele will face similarly high operation costs and, therefore, it could also require government subsidies.

A third issue, directly relevant for credit insurance, rests on the reasons for loan defaults faced by the banks. Loan default may be due to low incomes, because failure of the marketing system which results in low products prices, or high input prices or lack of inputs, or due to moral risk. In those cases, few arguments can be made for the benefits of credit insurance. However, credit insurance would be useful if loan defaults are due to income instability because of production losses.

To conclude, agricultural risks make farmers income stochastic, decrease debt bearing capacity and hence result in low loan repayment. However, stabilizing farmers income is not a guarantee of loan repayment. If, on the other hand, credit insurance is demanded, it would guarantee loan repayment to the bank only when yield failure is the reason for poor loan collection. The effects of credit insurance on the supply of credit are examined later on in this chapter.

ADMINISTRATION OF THE LOAN PORTFOLIO

Introduction

It is evident that there are conflicts of interest and policy in agricultural development banking. These emerge from the philosophy of financing risky agricultural production at the lowest rates of interest. Furthermore, managing an ADB is highly exposed to political decisions that have important consequences for the bank's growth, and hence, for the supply of credit. Although managing an ADB goes beyond management of the loan portfolio, this point is discussed here because of its importance in the mechanism of the supply of credit over time.

This may be taken as the case of a rather specialized ADB. On the asset side the bank is assumed to issue only loans of different maturity, size, expected return, risk of return, and demand for bank's human and physical resources. On the liability side the bank could engage in short and long term borrowing from the Central Bank, from commercial banks (with government subsidy), and from international financial agencies. The bank may also receive direct government subsidies.

This rather simple composition of the balance sheet is typical of many specialized ADBs. This, however, takes an extreme position, because some banks have, on the asset side, cash, securities (bonds), real assets and some fixed assets. Also on the liability side, some ADBs handle demand and time deposits on some pending accounts and net capital.

The Loan Portfolio: Administration Cost and Recovery Rates

Because of the nature of ADBs, their loan portfolio is typified by high costs and low returns. High total costs are the result of a large number of loans that the bank has to issue and administer. Low returns are due to low nominal rates and high default or, in other words, poor loan collection performance. This section analyses the interrelations among these concepts.

The loan portfolio is structured by loans issued for different purposes and of different sizes, maturity and risks. Although other development banks finance mainly large projects, ADBs finance mostly production loans for short cycle crops and cattle fattening, and a more reduced number of loans for investments like orchards, cattle herds and farm improvements. Although a significant proportion of the resources is allocated to a small number of large loans, the bank still has to administer a large number of small loans. Because most of the small loans are for annual crop production, their maturity is of less than one year, regardless of the size of the loan. A small number of loans however, may be of longer maturities, particularly the investment loans. As the bank diversifies by regions and crop cycles, there is likely to be a difference in the risk characteristic of loans, yet this may not always be the case

An ADB issues loans of different characteristics in response to farmers' demand, the availability of financial resources and following government policies. The allocation of funds is rarely made with a simultaneous consideration of the availability of physical and human resources. As a result the bank has to administer loans beyond its capacity, which in turn contributes to poor loan supervision and hence increased delinquency rates. This overloading of the banks' administrative capacity has also been suggested as the main reason for a poor service and hence unsatisfaction of the farmer.

There are however, important relationships and trade offs between financial policies and bank resource availability, as discussed next. If we assume that a loan of size (amount) L is charged a nominal rate of interest r, then the interest earning is

$$I = L.r. \tag{1}$$

Therefore, at maturity the bank would collect:

$$L = \bar{L}(1+r) = \bar{L}+I. \tag{2}$$

In fact however, the bank works with expected values on loan collection, which implies that the net recovery i.e., the proportion collected of each loan at maturity is

$$F(\bar{L}) = \gamma[\bar{L}(1+r)] \tag{3}$$

where γ is the recovery rate. The above suggests that the bank can increase expected loan collection in two ways. First, by increasing the nominal rate of interest, which is fundamentally a political decision. Second, by improving the loan recovery rate, which is however, a managerial issue, and hence the bank can alter it by increased loan supervision or through credit insurance.

Most bank officers would agree that increased loan supervision provides increased loan recovery at a decreasing rate. Nevertheless, the nature of this response function would be different for various groups of farmers, the crops grown and the risks to which the farmer is exposed. It is clear also that to provide more loan supervision, within the available physical and human resources, the bank would have to decrease the number of loans issued. If the bank moves more towards private ownership (becomes more concerned with profit and least cost solutions) one can expect a decline in the number of small loan and the enforcement of stronger loan selection criteria. Also, if an ADB had the option, it could reject certain loans because historical experience shows them providing too small gross returns, or because of too high costs. Yet, as the bank leans more towards public ownership it is not likely to be able to practice much loan selection procedures, but to take those loans that are expected to fulfill government policies.

Lending costs have two important components. First, there is the cost of issuing and supervising a loan until maturity. Second, there is the cost of keeping on the books and prosecuting a loan that is overdue. These costs are expected to be the same regardless of the amount lent, although there could be small differences in costs, depending on the purpose of the loan and its maturity. Long term livestock loans for example, may have a larger issuance cost, but in general a low administration cost while the loan is outstanding. Administration costs could also be different for areas with different accessibility

Managing the loan potfolio is a central issue in agricultural development banking. The bank can improve its loan selection and supervision procedures and upgrade and enlarge its staff. Yet, there will be a limit on improvement on loan collection, beyond wich little can be gained by investments on improved management. If the reasons for default are the risks in agriculture, then income stabilization programs would provide benefits to the extent that farmers are willing to pay and the banks are able to collect. Alternatively, credit insurance could also be demanded when production risks are identified as an important cause of loan defaults.

Loan Recovery and Credit Insurance

Although much has been mentioned about goals in development banking, no specific criteria has been given yet for the institution's objective(s). However, banks in general, and ADBs are no exception, can be considered as utility maximizers in the sense that they trade risk for return. As such, the institution's board of directors would act as risk averse a la Baumol, i.e. the risk of return is given a certain weight in the decision-making. This would depend on forecasted situations and the current financial position of the institution. In other words, the institution can be assumed as maximizing a linear objective function with parameters as follows:

$$U = E(R) - \phi\sigma \qquad (4)$$

where:

U, is utility,
F(R), are expected returns over a multiperiod
horizon,
ϕ , is a constant risk aversion parameter,
σ , is the standard deviation of returns over a
multiperiod planning horizon.

In the extreme cases of a fully supported government guaranteed institution, would be zero, yet that may exist only in extreme situations. As risk aversion increases, the bank would prefer to invest in the most secure loans, i.e. those with the highest recovery rate. Hence, the opportunity cost of credit insurance is expected to rise as the banks become more concerned with risk management. In other words, credit insurance becomes more desirable when the ADBs depend more on their banking capacity, than on their bargaining ability to obtain government subsidies to cover up for loses. As the availability of government funds becomes more limited, ADBs will benefit from requesting farmers to take credit insurance. This would allow the banks to fulfill development goals, even without changing their internal management and financial policies.

This is not to say, however, that credit insurance is justifiable on all grounds; but that it could be considered as an alternative, if credit supply is to reach potential viable farmers exposed to risks in production. Such farmers are viable in terms of their average productivity. Therefore, credit insurance provides them with a guarantee of loan repayment when, for reasons beyond their control, they could not pay back their loans.

There is evidence that agricultural credit insurance provides direct benefits for the lending institution. The insurance agency pays the bank the farmers' debt when farmers income is reduced because of crop yield failure, animal death or loss of function. In Panama in 1979 and 1980, the Agricultural Insurance Institute (ISA) payed the Agricultural Development Bank (BDA) indemnities for US$194,642 and US$402,143 respectively, which allowed for a significant improvement on loan recovery(ISA, 1981). The loan recovery rate for industrial tomatoes for example, was improved from an average of 82 percent between 1976 and 1978, to 95 percent in 1979, and 99 percent in 1980 (Pomareda y Fuentes, 1981).

Other advantages of credit insurance to the lending institution could be the reduction of costs of "farmer hunting" to collect the delinquent loans, and the additional supervision for the most optimal use of credit. The insurance supervision program helps the bank to separate those farmers that do not want to pay from those that can not pay. For the latter group the insurance agency will pay the bank the amount due by the farmer. However, since credit insurance provides coverage only for yield losses, its protection is only partial, because farmers can still have reduced incomes because of excessive costs of production or low product prices. Credit insurance would therefore, provide the largest benefits for the bank when lack of loan repayment is due mostly to yield failure.

It should be pointed out that although credit insurance allows the bank to show a healthier loan portfolio, it could be interpreted as a cover up for the bank's low capacity to recover its loaned funds. In this sense, credit insurance does not offer an incentive for the bank to improve its loan selection procedures and inspection practices to increase loan recovery. However, it is a way of improving loan recovery and it should allow the bank to grow at a faster rate. Its desirability is clearly high for the bank, yet its demand by farmers will be a function of premiums. Furthermore, the justification of insurance is to be based on cost effectiveness; i.e. whether the overall costs to the bank and the insurer do not exceed the benefits.

An issue for debate still remains. If credit insurance compensates for losses in agricultural production and therefore stabilizes farm incomes, then there should be no reason anymore for such highly subsidized interest rates when the argument for their use is the high risk of agriculture. With credit insurance therefore, agricultural development banks could charge higher interest rates to farmers. The net effect on the bank would, therefore, be a higher recovery of loans and higher interest earnings. Under such scenario, should it be the farmers who pay the cost of insurance or should it be the bank who pays for it?

From the viewpoint of the farmer, who is used to paying a low price for credit, it is unlikely that he would be very willing to pay the cost of insurance and a higher interest rate. On the other hand, if credit insurance provides valuable direct immediate benefits for the bank, it may be reasonable to think that this institution should help to pay for the cost of insurance. This is an issue for further research; but to estimate the maximum benefits of insurance for the bank it is assumed in this work that the bank does not pay any of the insurance cost.

ASSET-LIABILITY MANAGEMENT AND CREDIT SUPPLY

The Determinants of Bank Behavior

This section examines decision making of the banking firm by focussing on the factors that determine bank behavior. The presentation relates the decision making framework to the elements of portfolio theory discussed later.

The behavior of the banking firm is determined by its objective or set of objectives, its available options or choices for sources and uses of funds and the restrictions imposed by technology, physical inputs, laws and/or regulations. Decision making within such a set of objectives, alternatives and restrictions is a complex process and serious attempts have been made to model it. The modeling efforts, although they are abstractions and have limitations, have also shown to be of practical use in helping to build a theory of the banking firm and assisting decision makers.

The decision unit within the organization, whether an individual or a board of directors, provides guidance and exercises control, yet it is not expected to have full domain of every operation. However, its decisions are expected to maximize the organization's objectives. Such objectives are related to management interest as well as to ownership ones. Furthermore, in development banks such objectives are definitely related to government policy. Defining the institutions' objective is a difficult task, in part because no single objective can be specified. It is perhaps more proper to think of a financial objective as the maximization of expected utility and a set of goals. Through the maximization of objectives, the institution pursues goals mainly in terms of size and growth. The latter suggests that decision making is a dynamic-constantly adjusting process. Yet, as it is common in economic theory, we begin with a static situation or a so called equilibrium situation.

The alternatives for sources and uses of bank funds are selected from a wide set of possibilities, following the institution's goals and within the governmental rules and regulations, state and federal laws, and banking agreements. Hence, given a set of assets and liabilities, the bank will choose among them trying to satisfy its objectives.

Banks behave within technical and legal constraints. The latter are imposed by the Central Bank authorities and the Bank Commissions. These constraints vary by country and by regions within a country, and sometimes according to particular characteristics of the banks. One of the main restrictions on bank portfolio behavior concerns deposit and reserve requirements. Legal restrictions affect bank behavior in relation to allowance to invest in common stock of non-bank enterprises, size of operations, payment of interest on demand balances, underwriting of corporate debt or equity instruments and levels of interest rates.

The constraints in general can have positive as well as negative effects on the bank's behavior. It has been suggested that "these regulations effectively prohibit a number of otherwise attractive portfolios, and therefore; tend to impede banks from maximizing their objective functions. On balance such regulations probably lower bank profits, strenghten the hand of ownership interest relative to management, and force banks to have portfolios that lessen the probability of bank failure" (Hester and Pierce, 1977; p.18).

Physical constraints are important for a bank as for any other firm. Objectives can be maximized only within the availability of inputs such as labor, building capacity, computer time, telephone services and, in the case of banks servicing the agricultural sector, the number of vehicles and field staff to make credit supervision. The opportunity cost (or shadow price) on these constraints are the best indication of the need to modify them, particularly when planning the firm's growth.

This introductory section has presented the elements for decision making in the banking firm. It can be inferred that knowledge of objectives, alternatives and constraints provides basis for modelling the bank's decision making process, following the principles of portfolio theory. This, in summary, provides the rationale for holdings of alternative assets and liabilities on the basis of their expected return, their variance of return and the covariance of returns among them.

Before closing this section two points deserve further emphasis.

The first refers to the fact that proper bank management goes beyond the structure of the bank portfolio. Jessup (1980) points out that the consideration of cash flows, time deposits, loan decisions and reviews, projecting growth, managing bank capital, and other issues lead into complex decision making for sound, efficient and professional bank management. However, bank portfolio size and composition, in order to satisfy goals within the existing set of financial and physical constraints, is a very important determinant of successful bank management.

The second point refers to the time dimension in bank decision-making and its implications for bank portfolio management. The time variable is of singular importance when planning bank growth, particularly when considering that the institution's goals go beyond annual profit maximization. It must be recognized that, because assets and liabilities have different maturities and there is a transfer of resources between time periods, the bank faces a basic problem of dynamic balance sheet management.

Managing the Asset-Liability Spread

Successful bank management rest strongly on the management of the spread between assets and liabilities. Since development banks have enjoyed a preferential treatment, the cost of funds has been low; thus allowing them to hold assets (loans) of relatively low profitability. Yet, as the cost of funds becomes increasingly high, there seems to be no other alternative but to inject more profesionalism in the management of bank funds. This may allow the bank to continue providing development services at the lowest possible cost.

For simplification purposes, the analysis in this section abstracts from the other two important aspects of bank management, liquidity and risk, to concentrate on return i.e. the spread between assets and liabilities. With this purpose in mind, this section analyzes the cost and the revenue components of net earnings.

Financial costs in period t, $(FC)_t$, are defined by the interest and amortization payments on the banks contractual debts and interest payments on deposits from the public.

$$(FC)_t = \sum_k (b_k B_k)_t + \sum_k (A_{kt}) + (d_t D_t) \qquad (5.a)$$

where:

b_k , is the nominal rate on borrowed funds from source k.

$(b_k B_k)_t$, is interest expense due to source k in period t.

A_{kt} , is the amortization of borrowed funds from source k, due in period t.

d_t , is the interest rate paid on deposits in period t.

D_t , is the volume of money held on the form of deposits.

With respect to borrowings there is no element of risk, as the bank arranges for repayment conditions in advance. The bank would choose from alternative sources of funds not only as a function of the interest rate on borrowed funds, but the length of the repayment period and the grace period, a typical feature of development funds. As far as time deposits, the risk of withdrawal will be expressed through a liquidity balance constraint.

Operating costs are a major determinant of bank net earnings. Furthermore, there are important trade offs between the availability and use of operating capital and resources and bank performance. Operating costs include costs of personnel for loan appraisal and supervision, and maintenance of deposit accounts, vehicles for field work, maintenance of office facilities, computer services, etc. In the case of an agricultural development bank, operating costs (OC) , would include the following components:

$$(OC)_t = P_t + V_t + O_t \qquad (5.b)$$

where:

P_t = personnel costs
V_t = vehicles maintenance and repair
O_t = other expenses including office facilities and other

The bank can increase the loan recovery rates through stronger loan selection procedures, hence rejecting those loans that do not qualify; or else rationing credit according to debt bearing capacity criteria. Doing this, however, implies a larger amount of time spent on each loan appraisal, and hence, a larger demand for bank staff. Similarly, loan recovery rates can be increased through more intensive loan supervision, which implies more continuous contact between the loan officer and the farmer; therefore, larger requirements of staff, more vehicles and more operating capital for vehicle fuel, maintenance and repairs.

In trying to save operating costs, the banks contribute to increased loan defaults. Yet, because of the rather large number of loans administered, the bank has no choice but to accept the operating conditions, and hence the resulting performance. For better operating conditions most agricultural development banks would have to increase and upgrade their staff, and increase considerably their operating budget, in order to minimize default rates.

With respect to the generation of income, the bank usually has the choice of loans, investments and real assets plus any cash holdings. Returns on these assets will determine the banks' gross earnings, (GR), as follows:

$$\sum_i N_{i,t-j} L_{i,t-j} + \sum_i [N_{i,t-j} r_i L_{i,t-j}] \gamma_i + \sum_i M_{i,t-j} (1+y_i) Y_{i,t-j} + S_t + C_t$$

where:

$N_{i,t-j}$, the number of loans of typle i maturing in year t and issue in year t-j, i.e. j is the period to maturity.

$L_{i,t-j}$, the size of a loan of type i, maturing in year t and issued in year t-j.

$M_{i,t-j}$, the number of bonds of type i maturing in year t and issued in year t-j.

Y_i , return on bonds of type i.

$Y_{i,t-j}$, principal of bonds at date of purchase in year t-j.

S_t , returns on real assets

C_t^t , cash holdings

Therefore the bank's annual net return is determined by:

$$\pi_t = (GR)_t - (FC)_t - (OC)_t . \qquad (6)$$

The conflict of purposes in the management of agricultural development banks can now be more fully understood. On one hand, for development purposes within currently conceived philosophies, ADBs want to reach the largest number of farmers and provide them with the lowest price credit. On the other hand, the banks face high operating costs, low earnings, and very slim operating budgets. The result can not be other than a poorly performing institution characterized by high loan default and low quality credit supplied at low interest rate, but at a high cost to the farmer.

Dynamic Balance Sheet and the Supply of Credit

Bank portfolio management relies fundamentally on dynamic management of the balance sheet, i.e. decisions regarding the sources and uses of funds over time. This section develops the principles for the determination of the supply of loanable funds for the case of a specialized ADB.

The amount of funds that the bank can allocate in period t to loans of different characteristics, is limited by the availability of loanable funds:

$$[\text{loanable funds}] \leq [\text{loan recovery}] + [\text{bond collections}] + [\text{net deposits}]^{10} +$$

$$[\text{borrowings}] - [\begin{smallmatrix}\text{financial}\\\text{costs}\end{smallmatrix}] - [\begin{smallmatrix}\text{operating}\\\text{costs}\end{smallmatrix}] - [\begin{smallmatrix}\text{other}\\\text{investment}\\\text{securities and}\\\text{bonds}\end{smallmatrix}]$$

$$-[\text{cash}]$$

$$\sum_i N_{i,t} L_{i,t} \leq [\sum_i N_{i,t-j} L_{i,t-j} (1+r_i)]\gamma_i + \sum M_{i,t-j} Y_{i,t-j}(1+y_i) +$$

$$[(TD)_t + (DD)_t] + (B_k)_t - (FC)_t - (OC)_t - \sum_i M_{i,t} Y_{i,t} - C_t.$$

Where all terms have been already defined.

Loan recovery in period t is a function of management and financial policies in previous periods, which directly or indirectly affect expected returns through the nominal rate of interest or through the loan recovery rate. Furthermore, time preference and resource availability will affect the issuance of loans maturing in period t, t+1, t+2... t+j. With fixed low interest rates and high inflation the bank would prefer shorter loans; however, longer maturity loans demand less fixed issuance cost per period because money is turned over less frequently. Therefore, when trying to optimize its resources, the bank faces a trade off between the value of money, the operating costs and the availability of physical resources.

Borrowings in period t are not exogenously determined. The bank is not likely to be able to borrow beyond its financial and administrative capacity, as appraised by the bank's authorities and the lending agencies, either domestic or foreign. In addition, the leverage requirements and the overall composition of assets and liabilities will determine the optimum amount of borrowings at a given interest rate and repayment conditions. However, we could specify that there is always some upper limit on the amount of total borrowings.[11]

The bank would, therefore, select its sources of funds simultaneously with the decision on uses of funds, searching for the largest earnings margin, while fulfilling its development goals and servicing a particular clientele. In many cases, however, the bank negotiates the repayment conditions on borrowed funds depending upon the use to which the funds are to be put.

Financial costs in period t are determined by contractual arrangements in previous periods. Operating costs are defined in the previous section and they are not to exceed the institution's budget.

This analysis provides the rationale for bank growth and the supply of credit in each time period.

It is evident that interest rate policies, borrowing strategies, management of inflation; and allocation of physical and human resources are important determinants of bank growth. Simultaneous decisions on these issues will determine the bank's capacity to supply larger amounts of credit.

Differences in profitability, risk, maturity and resource requirements, will determine the allocation of funds to particular loans. These would include various annual crops, perennial crops, livestock and farm improvements. However, the bank may be limited in its decisions by government legislation, institutional agreements and/or political pressure to guarantee the supply of credit for particular purposes. These restrictions would affect the bank's optimal resource use and aggregate credit supply; but they would, on the other hand, guarantee short term fulfillment of development goals. One could, with certainty, indicate that the nature of these restrictions provides the fundamental difference between the operating practices among commercial and development banks.

The central issue and main conclusion of this discussion is that an ADB that depends primarily on loan collections, can not grow as fast as a diversified bank; hence it needs government subsidies and or external low cost funds. However, for an ADB with most agencies located in rural areas, the point remains as to how could it attract time and demand deposits from rural residents-farmers and non farmers. The issue has been discussed in the literature of rural finance, but not as much as the aspects of subsidized interest rates and related topics. One explanation for the neglect of savings mobilization may be its inconsistency with policies of low interest rates on loans (Vogel, 1981) and the high costs of administering a large number of savings and checking accounts. In addition one must recognize that managing a bank with multiple functions is far more complicated than managing a lending bank which relies on donors' money.

For a bank to be able to attract time and demand deposits in rural areas, it has to offer an attractive interest rate on savings accounts. Without this, the rural population would utilize its traditional means of hedging against inflation (Vogel, 1981; Buser, 1976). Nevertheless, it should be recognized that in many developing countries were money markets are quite segmented and where there is money illusion, savings do take place at negative real rates.

SET OF HYPOTHESES

More financial resources are needed to allow for
faster and equitable agricultural growth. To achieve
such objectives, agricultural development banks are
expected to play a more meaningful role by increasing
the supply of credit. In achieving their targets,
however, ADBs face problems of institutional design
and policies that limit their performing ability. On
the other hand, lending to agriculture continues to
be a risky and costly enterprise.

This chapter has shown that ADBs could improve
their overall performance, by functioning more like
banks, while they fulfill development objectives.
ADBs could diversify their asset and liability portfolios and this should increase their availability of
funds and their capacity to act in a financially
unstable world. This would be mandatory in cases
when the governments can not provide more subsidies
and when foreign soft loans dry out. Nevertheless,
even with the best diversified and managed portfolios, the banks could still face significant default on agricultural loans because of production
risks. Hence credit insurance should be considered,
but not as a substitute for better management of the
loan portfolio.

The main purpose of this book is to present a
theory and evidence about the impact of alternative
policies on the supply of credit. However, because
agricultural development banks are rather complex
institutions, the various policies would have many
other effects besides the changes in the supply of
credit. Therefore, the analysis carried does not
focus only on credit supply but also on various
elements of importance in the decision making process,
for selecting among alternative strategies. Furthermore, some important changes on the credit supply may
be the result of external factors rather than policies controllable by the bank or by the government

With this observations, the following hypotheses
were tested:

a. Increased risk aversion on the bank's management, implying more concern for loan recovery, will
have a positive impact on long term growth. However,
the bank may be induced to reject loans that would
have otherwise been part of its 'development oriented'
portfolio.

b. Credit insurance provides benefits for the
bank through reduced costs of loan administration,
higher average recovery and lower variance of loan
recovery. These factors contribute to a larger availability of loanable funds and faster bank growth; but
there is a level of interest rates on loans high

enough to produce benefits comparable to those resulting from insurance.

c. Serving small farmers imposes a constraint on
bank growth through higher administration costs and
lower availability of funds. Hence relieving the
agricultural development banks from this requirement
will increased bank growth.

d. Higher cost of funds borrowed from commercial
banks and a decline in government subsidies will have
a negative impact on the bank's growth and its capacity to serve small farm agriculture. The impact is
aggravated by the fact that the bank does not have a
diversified portfolio on assets and liabilities.

e. Transforming a specialized bank into one with
multiple functions will allow the bank to increase its
size and the availability of funds. However, the
growth of the supply of credit to agriculture could
be affected by the relative profitability of loans
versus other instruments and by the cost of funds.

These hypotheses were tested through analysis of
sample information on loans issued in various years
and by using a mathematical programming model of bank
growth. Although these data came from only one bank,
some important implications can be addressed for
specialized agricultural development banks in general,
or in developing countries in other regions

NOTES

1. Public (private) banks are those in which all
of the capital stock is owned within the public
(private) sector and operating policy is under public
(private) sector control.

2. Poor loan collection is strongly influenced by
political decisions when the government wants to benefit particular groups who claim crop disasters.

3. Members of the Latin American Association of
Development Finance Institutions (ALIDF).

4. Loans of different size, maturity, risk and
physical resource requirements.

5. Many times the reasons the ADB gives for not
providing a loan are the absence of ownership title,
a plot that is too small or a technology adoption
capacity that is unreliable.

6. See Hanson and Thompson (1981) for a discussion and a simulation of farm debt bearing capacity
as affected by income and variability of income.

7. Von Pischke and Adams (1980) highlighted the
primary implications of money fungibility in agricultural finance, particularly as it makes difficult to

evaluate credit programs, where credit is usually treated as another production input.

8. Ladman and Tinnermeier (1981) present an interesting description of such capital markets in Bolivia, where money borrowed for agricultural purposes is used in construction and other sectors, where returns are much higher.

9. These may include farm loans issued by different banks.

10. The concept of net deposits implies the available funds after meeting the leverage and reserve requirements.

11. The amount borrowed is also affected by the possibilities for attracting deposits instead of acquiring debt.

2
A Multiperiod Model
for an Agricultural Development
Bank

INTRODUCTION

The preceding discussion illustrates the nature of decision making for an ADB. Managing the portfolio of a bank is a complex process, yet possible to model in the context of a mathematical programming formulation. This is because the bank has an objective, alternatives and constraints. Furthermore, it is possible to model this decision making as a multiperiod system taking into account the intertemporal linkages necessary to account for dynamic balance sheet approach.

Portfolio theory offers the appropriate methodology for a quantitative framework that allows to structure the interrelations discussed in the previous chapter. It also allows to integrate the principles of risk management and resource allocation for optimal composition of the bank's portfolio.

Portfolio theory and commercial bank portfolio models are reviewed in the first part of the chapter. The remaining of the chapter develops a model for an agricultural development bank. The approach begins with the building blocks of an ADB model and then develops the multiperiod formulation. This is done through the use of a multiperiod linear programming model.

PORTFOLIO THEORY AND THE BANKING FIRM

Background

The general principles of portfolio theory are built on the concept that investors face a utility function of money. They would choose among alternatives on the basis of their expected returns and variance of expected return.

Whether the options are bonds held by a private investor, loans issued by a bank, policies offered by an insurance agency, or crops grown by a farmer, these options are subject to risk because their returns cannot be anticipated with perfect certainty. This implies the existence of an expected return associated with a probability distribution and hence a variance or returns. Moreover, nature does not affect the behavior of returns for each alternative in the same way; hence, there is some degree of correlation in the returns of the various alternatives. To the extent that this correlation is not perfectly positive, there are gains from diversification.

The objectives of enterprises are multiple. From a financial point of view however, two are common to all investors: a) to obtain the maximum return[1]; and b) this return to be dependable and stable i.e., subject to the least variation. It is on these two objectives that one finds an inconsistency i.e., the maximum return is not necessarily the one with the least uncertainty of return. The choice among pairs is a function of the investor's preferences; yet it is necessary to know which are the portfolios that provide a given return with the least uncertainty.

It has become established now to measure risk by the variance of returns. However, in portfolio analysis, the covariance of returns of alternative investments is as important as the variance of individual investments. In fact, it is because of the existence of such a covariance that risk can be managed through diversification. The general principles of portfolio theory go back to the origins of decision sciences, but it was only with the pioneering work of Markowitz (1952) that a mathematic formulation was made available.[2] Ever since its appearance, portfolio theory claimed interest among economists and financial analysts. The following two sections present the concepts of feasible and efficient portfolios and then the concepts of utility and optimal portfolios. The separate discussion of these concepts is necessary for a fuller understanding of portfolio management theory.

Feasible and Efficient Portfolios

A portfolio can be structured by one or by many investments,[3] each with its own characteristics. The selection of the best combination of investments given the investor's capital constraints is not an easy task. According to Markowitz, the choice is among those investments for which the maximum return is associated with a given variance of return. The locus of such combination defines a set of efficient portfolios; i.e, the frontier in Figure 3.1. The combination of investments that for the same variance provides with smaller returns are located in the shaded area and they are defined as feasible but not efficient portfolios.

If the investor can allocate his financial resources into n investments, then the proportion of each one in the portfolio can be defined as X_i (for i=1,2..n)

$$\sum_{i}^{n} X_i = 1 \qquad (9)$$

and the feasible portfolios (including the efficient ones) fulfill the following non-negative conditions:

$$X_i \geq 0 \qquad (10)$$

The expected return of the portfolio, a random variable in itself, is the weighted average of the expected returns of each investment ($\bar{R}$) hence

$$E(R) = \sum_{i}^{n} X_i \bar{R}_i \qquad (11)$$

The covariance of return of the portfolio depends on the proportion of each component, its variance of return and the covariance of returns among investments.

$$\sigma_r^2 = \sum_i \sum_j X_i X_j \sigma_{ij} \qquad (12.a)$$

$$\sigma_r^2 = \sum_i \sum_j X_i X_j \rho_{ij} \sigma_i \sigma_j$$

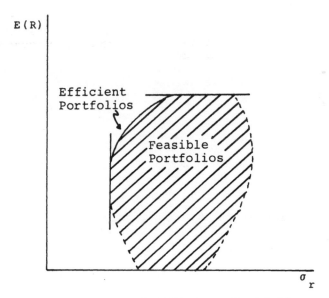

Figure 3.1   Feasible and Efficient Portfolios

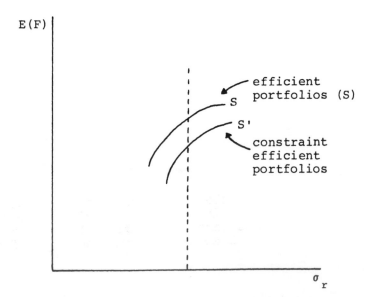

Figure 3.2   Efficient and Constrained Efficient Portfolios.

where ρ_{ij} stands for the correlation coefficient among investments and σ_i and σ_j are the standard deviations of returns. In matrix notation the variance of return is:

$$\sigma_r^2 = [X_1 \ X_2 \ldots \ldots X_n] \cdot \begin{bmatrix} \sigma_{11} & \sigma_{12} \cdots \sigma_{1n} \\ \sigma_{21} & \sigma_{22} & \sigma_{2n} \\ \cdot & \cdot & \cdot \\ \cdot & \cdot & \cdot \\ \cdot & \cdot & \cdot \\ \sigma_{n1} & \sigma_{n2} & \sigma_{nn} \end{bmatrix} \begin{bmatrix} X_1 \\ X_2 \\ \cdot \\ \cdot \\ \cdot \\ X_n \end{bmatrix} \quad (13)$$

The generalities presented above deserve further discussion to understand the role that each component $(X_{ij}, R_i, \sigma_i$ and $\rho_{ij})$ play in the performance of the portfolio.

The proportion of each investment, (X_i) contributes linearly to the expected return and variance of return of the portfolio. Therefore, larger values of X_i will have more weight in the expected return and variance of return of the portfolio.

In the special case in which the portfolio is composed entirely of one investment, correlation is of no relevance. However, in the general case correlation is important and this can be seen by further disaggregation of equation (12.a).

$$\sigma_r^2 = \sum_i X_i^2 \sigma_i^2 + \sum_{i \neq j} \sum_j X_i X_j \rho_{ij} \sigma_i \sigma_j . \quad (14)$$

The first term of the second part of equation (14) is obviously positive. However, if the covariance term is negative, it counter acts with the first and hence reduces the variance of the portfolio. The larger the degree of negative correlation in the portfolio, the larger the convexity of the efficient frontier.

It was Markowitz who first suggested quadratic programming to find the set of efficient portfolios.

The problem reduces to:

$$\text{Min} - \lambda \left(\sum_i^n X_i R_i \right) + \sum_i^n \sum_j^n X_i X_j \sigma_{ij}^2 \tag{15}$$

for all possible values of $\lambda \geq 0$ and

s.t

$$\sum_i X_i = 1. \tag{16}$$

In practice however, many other constraints can be imposed on the minimization condition. Even when assuming that all such other constraints exist, every value of λ will give a different efficient portfolio.

The concept of an efficient portfolio as discussed, relates strictly to financial criteria. In practice however, besides the risk-return criteria, there are a number of factors that limit the choice to a constrained set of efficient portfolios. Because of resource constraints, managerial criteria, laws and or regulations the investor may be limited to a set such as S' instead of S in Figure 3.2.

In the case of S', for the same amount of risk the investor can obtain a smaller return because of the existence of other constraints. Incorporating these constraints in the selection of the efficient portfolios is straightforward ,as the model in equations (15) and (16) can be further constrained by:

$$\sum_i a_{ij} X_i \leq A_j \tag{17}$$

where a_{ij} represent the resource requirement and A_j is a vector or resource constraints. Accounting for those constraints adds realism and practicality to portfolio models for individual or institutional investors who, besides a capital constraint, face a number of physical and institutional limitations.

Utility Maximization and Optimal Portfolios

Investors will choose from an infinite number of efficient portfolios, the combination that satisfies their own preferences. The latter concept implies the existence of a utility function i.e., some relation between risk and return, the two decision elements in portfolio theory. The investor would, under such conditions, try to maximize his utility function, which reflects his attitudes towards risk and return.

Because of the difficulty in estimating the nature of utility functions, much of the work on portfolio management has avoided this issue, although some economists have studied in detail the characteristics of particular utility functions and considered several criteria for their evaluation. The concepts of utility go back to the 18th Century with the work of Bernoulli, who hypothesized that investors maximize "moral expectation".[4]

It has become common to assume that most investors are risk averse i.e., they dislike risk, hence they face a concave utility function. One can assume this concave utility function to be a negative exponential, let's say

$$U(Y) = 1 - e^{-a} Y, \text{ when } a > 0 \qquad (18)$$

Freund (1956) has shown that the expected value of a negative exponential, integrated over a normal probability density function can be expressed as:

$$F[U(Y)] = F(Y) - \left(\frac{a}{2}\right) \sigma_y \qquad (19)$$

In practice, the estimation of a utility function faces severe problems. In general terms, however, we can assume the expected utility function in equation (19) expressed as a linear relation of $F(Y)$ and σ_y:

$$F(U(Y)) = F(Y) - \phi \sigma_y \qquad (20)$$

where ϕ is a constant risk aversion parameter. This development is due mainly to Tobin (1958), who explicitly distinguished between risk lovers and risk averters.[5] The first would be willing to trade a higher expected rate of return in exchange for a greater variance, while risk averters will only accept more risk if they can expect "relatively higher" gains.

A solution to the investor's choice of an optimal portfolio can be found at the tangency point of the efficient set and the highest utility function. Obtaining such an optimal solution requires maximizing the objective function (utility) subject to a set of constraints. Quadratic programming can be used to obtain such an equilibrium solution; however, linear programming also offers an alternative approach.[6]

The general maximization model that provides an optimal solution to this problem is

$$\text{Max } E\ (U) = F(Y) - \phi\sigma_y \tag{21}$$

$$\text{s.t.:} \qquad \sum_i X_i \leq F \tag{22}$$

$$\sum_i a_{ij}X_i \leq A_j \tag{23}$$

$$X_i \geq 0 \tag{24}$$

In this case however, the model is formulated with the X_i's being (absolute) amounts of money invested in each alternative and hence their total volume is not to exceed the available funds (F).

This general model can be applied to any financial institution and certainly to banks. The most important applications in the banking industry are reviewed next.

A REVIEW OF BANK PORTFOLIO MODELS

The Static Models

The behavior of the banking firm has been studied and modeled with considerable interest during the last two decades. There are a number of models and approaches which are being forged together to form a theory of bank behavior. Although econometric, simulation and optimization models are available, the concern in this work is only with the latter.

It is possible to analyze the theory and models used to explain the behavior of the banking firm in the context of their evolution with regard to particular issues. These issues are; first, the management of assets and liabilities; second, the incorporation of risk in the decision making process; and third, the simultaneous consideration of financial and real (physical) constraints in the determination of the firm's portfolio.

A large proportion of the literature on banking theory focusses on the analysis of partial static models of bank portfolio management. In these models

the size of the portfolio is assumed to be exogenously determined, while the aspects of resource needs
and constraints for the administration of the portfolio are ignored. These models focussed first on
the issue of reserve and liquidity ratio management.
Later on, with the influence of the contributions of
Markowitz (1959), the aspects of risk and its effects
on the composition of assets and liabilities were
introduced. The issue of risk management however,
becomes more relevant in the complete models of bank
portfolio management which are discussed later.

The basic models of reserve-liquidity management
are based on the work of Edgeworth (1888). The common feature of these models is that they treat the
reserve-liability decision as a problem of inventory
optimization under stochastic demand. It is assumed
that on the asset side the bank can choose between,
let's say, two types of assets: reserves (which
could be securities) and loans. The bank would
structure its portfolio trying to minimize the opportunity cost of excess holdings of the various assets.

In such a case, on the asset side, the problem is
one of choosing the optimal allocation of the given
funds among reserves and loans. This of course,
assumes that there is a minimum reserve requirement.
The bank usually has the opportunity to hold a
variety of assets and the choice among them is, as it
will be discussed below, a function of relative risk
and returns. In this regard, a priori selection of
customers can be done if realistic knowledge about
them is available. Therefore, it is hypothesized
that the bank, by spending resources on the collection of information about its customers, can reduce
the expected rate of default. Baltensperger (1972.a
and 1972.b) linked the cost of incomplete information
and the return to more information with the variability of certain random variables such as default loses
and deposit fluctuations.

With regard to management of liabilities,
Baltensperger (1980) indicates that the argument is
usually made that thebank does not have a choice
other than simply accepting all the deposits offered.
The end of the year net interest earnings provides
the bank with a new input in liabilities, known only
in some probabilistic way. Managing this flow of
earnings within the bank-capital account has also
resulted in interesting contributions by Baltensperger (1972.a, 1972.b) and Taggart and Greenbaum
(1978).

In summary, the static partial models of bank
management deal with questions of asset choice or
liability management. Their approach is comparable
to the principles of minimum cost production given a

level of output, according to the general micro-
economic theory of the firm.

In the context of this research, it is more relevant to focus on more complete models of the banking
firm, where the important issues are not limited to
those of asset-liability balance, but also to the influence of risk in management and the determination
of the firm's size. These more complete models
assume competitive behavior i.e., given interest
rates. There is however, a number of models that
incorporate monopolistic behavior in the banks' decision making and organization. The models developed
by Klein (1971) and Monti (1972) asume that the bank
has monopoly power and determine bank scale and portfolio structure via (net) revenue maximization.

These more complete models determine the optimal
structure of assets and liabilities (apart from the
exogenous capital account) as well as the total size
of the bank, under the assumption that it acts as a
price setter in the markets for bank loans as well as
the markets for different types of deposits. This
approach neglects resources constraints and it has
been criticized for not providing a rational explanation of bank behavior (Baltensperger, 1980).

Some of these models have taken risk into account
only in an indirect way. It is assumed that the firm
cares about risk only to the extent that it is
reflected in expected profit. The firm, therefore,
does not directly consider risk in portfolio management in the sense of trading between expected return
and variance of return. There are, however, a few
models which approach the theory of the banking firm
by applying the general theory of portfolio management under the assumption of risk averse behavior in
the bank's administration. This approach, based on
the works of Markowitz (1959, 1976), and Baumol
(1963) treats the banking firm simply as a collection
of financial assets with exogenous and stochastic
rates of return and with liabilities treated as negative assets.

Within this later group of models, the simplest
one is that developed by Pyle (1971). This model is
basically a decision tool for a financial intermediary which has the choice between three securities:
a riskless security and two securities with an uncertain yield over the given decision period, referred
to as "loans" and "deposits". The latter can, therefore, be held in negative amounts. The model is
rather simple, ignoring liquidity and solvency
considerations as well as resource costs and constraints. However, the interesting feature of the
model is that the firm maximizes a concave utility
function, i.e. it is risk averse.

Within the same framework, one finds the models

proposed by Parkin (1970), and Hart and Jafee (1974). Their models assume that the set of assets and liabilities which are held by the firm is institutionally given. This is a rather realistic assumption in light of segmentation in the money and capital markets (Van Horne, 1980). Both models treat the institutional aspects as exogenous constraints on the admitted range of assets and liabilities. A similar approach to treat this aspect is the one used by Kane and Malkiel (1965), also under the consideration of risk averse behavior, where utility is maximized over a one calendar period.

In general, these models treat satisfactorily the aspects of risk management within the principles of portfolio theory in a static framework. However, a complete theory of the banking firm must explain how the firm combines its scarse resources of various kinds in order to satisfy objectives within the portfolio management approach.. That is to say that the optimal composition of the portfolio depends not only on the risk-return trade-offs, but also it depends on the relative uses of the limited factors of production by each unit of the assets and liabilities. Within this framework, the literature on bank models is more limited; yet some valuable contributions have been made in the last years with regard to models that account for resource use.

The consideration of resource use in bank portfolio management was taken up first by Pesek (1970) with further refinements by Towey (1974), Adar, Agmon and Orgler, (1975), Sealey and Lindley (1977) and Saving (1977). These models essentially represent pure production cost models of banking i.e., they explain size and structure of bank liabilities and assets purely in terms of the flows of real resource costs. All other aspects referred to before, including risk management are not considered. Even the latest of these models, proposed by Baltensperger (1980, p.32-36) neglects specific account of risk management.

A known exception that incorporated risk averse behavior and accounted for constraints in the decision making, but in a static framework, is the work of Robison (1975) which is summarized in Robison and Barry (1977). Robison developed a quadratic programming model and used it to evaluate how an expected utility maximizing choice is changed in response to factors shifting the mean-variance (E-V) efficient set and changes in the decision maker's level of risk aversion. The author also reports consistency between bank behavior and the E-V decision criteria. The model was used to evaluate alternative agricultural finance policies on their effect on a commercial bank portfolio.

As noticed, the preceding discussion focussed on bank portfolio models within a static framework of analysis. Although all this work provides interesting contributions to the field of modelling bank behavior, it totally neglects the issue of time in modelling bank portfolios.

The Dynamic Models

There have been two alternative approaches to model bank behavior in a dynamic framework: simulation models and optimization models. While the first are useful to find out the effects of alternative decisions, they are of limited applicability in trying to determine what is the "best plan" for the bank in terms of the combination of its resources to attain specific objectives and goals over time. With this latter purpose in mind, optimization models offer a better alternative approach. They build on the principles of portfolio theory and some interesting applications have been made in the last twenty years.

The dynamic models of bank portfolios are perhaps some of the most interesting contributions in dynamic portfolio management. Given the operating procedures of banks, these models are more complete and useful than static models. In reviewing these models two groups can be identified. The models in the first group do not account for risk in the objective function, hence they are basically dynamic optimization models, with risk management policies built in the model constraints. These models are in the majority. The second are models that explicitly include risk in the objective function; hence the dynamic sheet balance optimization accounts for trade offs between expected returns and variance-covariance of returns among the alternatives for the sources and used of funds. Because of the complexities of treating risk in the dynamic decision making process of bank management, this latter category of models is known only on a theoretical basis.

The first known dynamic model is the one proposed by Chambers and Charnes twenty years ago. Their model finds the balanced asset portfolio which maximizes returns over a planning horizon, while meeting various constraints in each period. The "balanced" portfolio is the one that meets the regulations of the bank examiners of the Federal Reserve System. It is assumed that the banker knows the level that will prevail, at various dates in the future, of demand and time deposits, of rates of interest and the bank's net worth. The objective is to maximize profit and the investment choice is among loans, government securities and bonds, each with alternative maturities.

Chambers and Charnes' model accounts for risk only indirectly through leverage requirements as a function of the assets' liquidity.[7] Their major contribution can be found in the interpretation of the dual to the LP problem. This allows to measure the marginal rate of return on additional available capital at any time period, the opportunity cost of reserve requirements and the benefits of rediscounting and lending out.

As Cohen and Hammer (1966, p.89) have commented, the fact that the Chambers and Charnes approach employs a multiperiod model should not be misinterpreted to imply that the plans formulated for the entire horizon should be inviolable. The main purpose of the model is to help determine what actions should be taken at present (i.e., in the first time period). More than one period is hence included to avoid the suboptimal decisions which might result from a shortsighted view that ignores relevant aspects of the future. Still, the choice of the length of the planning horizon is a difficult task, usually depending on the stability of the environment and the length of calendar time represented in each time period in the model.[8]

A further refinement of the original model was offered by Cohen and Hammer (1967), including three alternative objective functions: 1) maximize the value of stock holder equity at the end of the final period; 2) maximize the present value of the net income stream over the planning period or finally; 3) maximize the present value of the income stream during the planning period, plus the stockholder equity at the end. The authors preferred this later function. In addition their model takes into account operational constraints as well as projections of various intertemporal relationships between stocks and flows of deposits, reserves and loan demand.

Cohen and Thore (1970) extended the model by Cohen and Hammer to include information contained in discrete probability functions that attach to levels of liabilities and interest rates. The problem then, became one of two-stage programming under uncertainty. In the first stage, the interest rates are determined for future time periods and in the second stage, they are taken into account in making current decisions. Further developments of the two-stage programming model were proposed by Crane (1971). Crane's model maximized the expected rate of return, using two-stage programming under uncertainty subject to some multiperiod inventory constraints, to assure that purchases did not exceed net cash flows plus sales. Although Crane's model did not include risk in the objective function, it included a constraint to limit the maximum capital loss. Because the model included very few constraints, the results of the model were unrealistic since only one or two assets were considered.[9]

The models reviewed above were basically normative. Their intent was to produce solutions that reflected optimal conditions and hence suggest changes in the current portfolio moving towards optimality. However, the lack of realism of those models could have provided "optimal" solutions that were impractical, institutionally infeasible or politically unacceptable.

Other dynamic portfolio models have been designed with the purpose of explaining the current portfolio and its adjustment process. By removing the current constraints or by changing specific parameters one could asses the effects on the portfolio structure and its new adjustment path. This can be described as a choice and adjustment process. It is a choice in the sense that the model selects an optimal equilibrium portfolio while an adjustment would involve the speed and method of moving towards some new equilibrium.

Among these models the first one is the purely theoretical case proposed by Porter (1961). The model included risk explicitly, in the objective function, as this was the maximization of utility and not just the maximization of profit. Other models were developed within this line of work, such as those of Fried (1971), Hester and Pierce (1975), Frey (1977), and Beazer (1975).

Explicit account of risk in a dynamic model's objective function is cumbersome and the issue has been avoided. If the model was to maximize a multi-period utility function it would be possible to write

$$U = \sum_{t=0}^{T} E(R)_t - \phi_t \sum_{t=1}^{T} V(R)_t \qquad (25)$$

where, assuming that activities can be designated by X_i, then

$$\sum_{t=0}^{T} (R)_t = \sum_{t=0}^{T} \sum_{i=1}^{I} R_{it} X_i \lambda^t \qquad (26)$$

where R_{it} is the return and λ_t is the appropriate discount factor for a given price of capital (ζ),

$$\lambda_t = \frac{1}{(1 + \zeta)^t} \qquad (27)$$

The variance of present value, however, is calculated as a function of the variances of each period's present value of cash flows on the net available funds (see Anderson, Dillon and Hardaker, 1977, p. 260).

To do this it is necessary to take account of the covariance between the cash flows of different time periods. The relationship is

$$V(R) = \sum_{t=0}^{T} V[(R)_t] + 2 \sum_{t'=0}^{T-1} \sum_{t=t-1}^{T} cov[(R)_{t'}, (R)_t] \quad (28)$$

$$V(R) = \sum_{t=0}^{T} V[(R)_t] + 2 \sum_{t'=0}^{T-1} \sum_{t=t'-1}^{T} \rho_{t',t} \, \sigma(R)_{t'}, \sigma(R)_t$$

(28.a)

The covariance term is of particular relevance in this case as funds available in period t are a function of funds available in all previous periods. Similarly the risk in fund availability in period t is related to the risk in previous periods.

As the incorporation of this covariance effect would add significant complexities to the model, alternative more simple approaches have been used. Fried (1977) utilizes chance-constrained programming to accommodate the Markowitz formulation to stochastic movements in deposits levels and returns and to provide trade-offs between risk and return. The model maximizes expected return subject to probabilistic constraints on acceptable levels or risk and illiquidity. The risk constraint is formulated as:

$$E - K \sigma \leq \epsilon \quad (29)$$

where E is the expected return on the portfolio, σ is the standard deviation, K is a number of standard deviations, and ϵ is a safety level or lower limit that the expected return must exceed with certain probability. The solutions were compared with the actual portfolios chosen by the banks and the author concluded that the actual bank portfolios were inefficient, since they had rates of return lower than the ones selected by the model. This assertion however, assumes that the model formulation actually represented the bank's real objective and all constraints; which could have not been the case, as institutional and resource constraints were not included.

Hester and Pierce (1975) developed a multiperiod LP model to address the issues of growth, size and market imperfections created by public policy and unforseen events. Although the choice issue in bank portfolio models is very important, Hester and Pierce (1975) emphasized more the adjustment process as a result of changes in policies. They developed a dynamic model that allowed to determine that the length of time for the adjustment process to be completed, was approximately eight weeks for cash inflows and eight months for mortgages. Their analysis of the adjustment of bank portfolio was of particular relevance for the lagged effects of monetary policy. The authors concluded that costs of adjustments to external shocks and variations in interest rates are very important in accounting for variations in bank portfolios and profits.

One of the most complete dynamic bank portfolio models is the one developed by Beazer (1975). The aim of this work was to compare the portfolio chosen by the model with the one actually chosen by the bank. The author insisted that the bank's criteria is not profit maximization but utility maximization. The bank's objective and choice function involved not only a trade off between risk and return, but also included liquidity considerations. It maximized the rate of return of the portfolio but warned that such an objective may not lead to the same results as maximizing the present value of the expected earnings stream. However, this was operationally more simple and "probably more closely approximated what (bank) portfolio managers actually use as a criterion function" (Beazer, 1975; p. 23). The model was described as a portfolio model in which the choice set was defined by the trade offs between risk and return. But the risk measures in the model were not in the objective function but in the form of liquidity constraints. In this sense, it was an unconstrained feasible set; however, because of the capital adequacy ratio, required reserves and other constraints, the choice set was reduced to a constrained feasible set.

In spite of the significant contribution of Beazer's model, it was still a financial management model which did not take into account physical constraints. It was assumed that if a portfolio was financially sound, then the bank will adjust its physical needs accordingly. Although a valid assumption in large commercial profit oriented banks, it is questionable in the formulation of short term policies among small commercial banks and among development banks which may have severe limitations for immediate adjustment in physical resource needs. Hence, in the later case such constraints must be considered, therefore, defining an even narrower feasible set.

Concluding Comments

This review is indicative of abundant research on bank portfolio models. Yet, the work is a disseminated set of bits and pieces that have not been linket together in a generalized theory of bank portfolio management. Furthermore, all the work is within the area of commercial banking in developed

countries. However, in developing countries, political constraints, risk and (more) imperfect capital
markets add complexities to the management of the
bank portfolio. Nevertheless, the existing models
provide valuable insights that can be adapted to the
special case of agricultural development banks.

THE BUILDING BLOCKS OF AN ADB MODEL

Building a portfolio model for a specialized
agricultural development bank has the purpose of
evaluating the current practices and restrictions and
exploring alternative means for a more rapid and
stable bank growth. From Chapter 2 it is clear that
specialized agricultural development banks operate in
an environment where institutional design and policy
limit the sources and uses of funds. Also, socio-
political factors impose additional constraints that
affect the financial performance of the institution

From the previous sections in this chapter, it
can be inferred that while a number of commercial
bank portfolio models have been built with the purpose of addressing specific issues, none of these
models provides a quantitative framework for agricultural development bank portfolio management. Also,
there are not portfolio models for development banks.
However, by building on several aspects provided by
earlier commercial bank models, it is possible to
propose a portfolio management model for an agricultural development bank

The model conceptualized here builds on earlier
efforts, and hence, it resembles the treatment of
assets and liabilities in the pioneering work of
Chambers and Charnes (1961) and Beazer (1975); it
treats discounted flows of money as suggested by
Cohen and Hammer (1967); it takes into account resource constraints as suggested by Baltensperger
(1980) and builds on the dynamic framework used by
Fried (1970). Further refinements are introduced,
including disaggregation of the loan portfolio for
different crops and livestock; constraints that
reflect development bank goals; a measure of risk
that enters directly the ban 's objective function;
and a general formulation that allows for the
simulation of alternative relevant policies. These
policies include credit insurance, alternative interest rates, changing the clientele, and ultimately
modifying the bank's institutional constraints.
Furthermore, the formulation allows, with relatively
few changes, a transformation of the model for a
specialized bank into one with multiple functions,
including those of handling savings and checking
accounts and other liabilities.

To model the bank's decision making framework and to accommodate the needs of the analysis, a multi-period linear programming model has been chosen as the appropriate quantitative tool. The mathematical programming model is structured with three basic components: i) A set of alternatives which in this case are the various sources and uses of funds or, in the context of the bank's balance sheet, the assets and liabilities. ii) A set of restrictions (constraints) which include the availability of resources, the financial and operating contraints, and the institutional (political) restrictions. iii) An objective (the objective function) to be optimized, which in this case is multiperiod utility to be maximized.

As it will be discussed later, the objective is utility in a multiperiod horizon, hence the model includes a measure of risk. Also, several constraints and activities are needed to allow for inter-temporal transfer of funds. The alternative sources and uses of funds, the institutional and physical resource constraints, and the financial constraints are discussed in the following sections.

Alternative Sources and Uses of Funds

There is no reason why a development bank could not fulfill the functions of a financial intermediary. In such a case, on the liability side the bank could accept time and demand deposits, borrow from the Central Bank and from other national or international banks, and hold equity. In the asset side, the bank could issue different types of loans and mortgages, and invest in alternative securities.

The general model is developed for a development bank which can fulfill all the functions of a financial intermediary, but the application in the next chapter is to the case of a specialized bank. Therefore, on the asset side (the uses of funds) the alternatives are the issuance of loans, the acquisition of investments and the holdings of cash.

The loans can be classified in terms of their use for annual crop production, establishment of orchards, herd build up, cattle fattening, farm equipment and machinery and farm improvements. Loans for annual crop production and cattle fattening are of less than one year of maturity. The other loans have maturity periods between two and five years. Annual crop production loans usuallly represent at least fifty percent of the number of loans issued, however; in terms of value, they may account for only thirty or forty percent of the loan portfolio.

Because of the clientele the bank serves, loans

should be properly disaggregated by sizes. This is necessary because the bank usually imposes requirements of serving a large number of small farmers with the consequent high costs that such practice imposes. Another criteria on the disaggregation of loan categories is whether they are insured or not. As it has been discussed before, insured loans provide higher average returns to the banks. However there could be restrictions on their number and on the volume of insured credit because of the capacity of the agricultural insurers and/or because certain loans do not meet the insurer's selection criteria.

If we define the principal of a loan as L, then we can distinguish the following categories:

$$L_{ijkm}$$

where:

i = purpose   (i= 1 --- I)
j = 1 if insured, zero otherwise
k = size (k= 1 --- K)
m = maturity (m= 1 --- M)

The other investments or options to the bank include government and agency securities of various maturities and stocks. The purchase of securities in this case is restricted to government bonds of maturities in annual intervals. Shorther term maturities may be preferable to take advantage of cyclical availability of funds influenced by the seasonal demand for loans. Yet this can not be handled with an annual time structure of the model. It is also assumed that securities can be redeemed only at maturity.The legislation in each country may inhibit the bank from investing in particular types of securities. Holdings of Cash and reserves should also be specified as a proportion of the bank's total assets.

In summary the bank's choice of assets include: Loans (L_{ijkm}),government securities (G_m), and cash (C). Except for cash, the bank could hold any number of these assets, with a wide range of nominal values and other characteristics. Hence the decisions concern the number of each type of asset to acquire.

The liability options for a development bank are, as for most banks, demand deposits, time deposits, borrowings from the central bank, from commercial domestic banks, from commercial international banks and from international development agencies. Agricultural development banks also receive significant amounts of government subsidies. ADBs rest primarily on this last source in the form of allocations from the central banks (or from the treasury) which obey macroeconomic policy and agricultural development strategies. Although, this source of funds has been the primary one for most development banks, its future continuous supply is jeopardized by the difficult financial situation of most developing countries. Paradoxically, the banks have obtained larger amounts of these resources when their performance has been worst. Such allocations have been provided to allow the banks to continue in operation, and fulfill their "development" role in spite of their financial performance.

Demand and time deposits have not been a primary source of funds for ADBs, yet the diversification of the banks to take advantage of these opportunities is largely advocated. The issuance of savings accounts demands the offering of a competitive rate. It is assumed that at such rate the bank would attract deposits as to optimize its portfolio, and hence an unlimited amount of potential customers.

Borrowings from commercial banks (domestic and foreign) is also practiced at competitive rates, therefore; this activity requires government subsidies. There is in addition a complexity that arises because of foreign exchange risk on funds borrowed in the international market. This, however, is not considered here.

Borrowing from international agencies at low interest and other easy conditions has been a major source of funds, particularly during the last two decades. That, however has taken the countries into severe indebtedness. As a result, it is not likely that the rate of growth of such funds could be maintained, unless the banks show a higher ability to administer such funds. Alternatively, the development banks would borrow more significant amounts from commercial banks.

In the case of liabilities, the model allows for different sizes of checking accounts (as demand deposits) and savings accounts (time deposits) but borrowings and subsidies are specified as total amounts of money. In summary the choice of liabilities include: demand deposits (DD), time deposits (TD), borrowings from commercial banks (BC), borrowings from international development financial agencies (BI), and government subsidies (GS).

Purposely this section has begun with a description of the bank's alternatives in the sources and uses of funds, to provide the basis for the following discussion of how the bank could choose from among those alternatives, within the limits of its operating constraints.

Institutional and Resource Constraints

The bank's authorities, (following the principles of portfolio theory), could choose from among efficient asset and liability portfolios contained in an unconstrained set of feasible financial solutions. Yet, in reality the choice is limited to a more restricted set, because of the existence of institutional and physical constraints.

The institutional constraints are those that are established to fulfill governmental policy and those that obey current bank organization. The spatial allocation of funds is also in some cases an important institutional constraint. These could also require balance of small and large loans and the assignment of volumes of credit for particular crops.

If restrictions existed on minimum amount of funds to be allocated to particular crops or livestock, this constraint can be specified as:

$$\sum_i \sum_j \sum_k \sum_m l^t_{ijkm} L^t_{ijkm} \geq L^{t*}_{j=1} \qquad (30)$$

where:

l^t_{ijkm} is the number of loans of type $ijkm$ issued at time t.

L^t_{ijkm} is the principal of loan of tyle $ijkm$.

L^{t*}_i is the minimum amount of funds that must be assigned to purpose i in year t.

This is a particular situation for export crops that provide the basis for the country's foreign exchanges earnings or food grains to provide basic food supplies. These constraints could alternatively be formulated in terms of required volumes of production and let the bank allocate the funds in the most optimal financial way given costs of production and yields.

Another important constraint may be defined by the value of insured loans that can be issued every year. In addition, there could also be restrictions on insurance to be provided for only certain crops and regions. The above constraints are not defined by the bank authorities, but by the insurance agency according to financial and resource availability constraints, geographic coverage or particular rules and regulations that exclude certain bank loans from the insurable set of loans. If, out of the total loans issued by the bank, we can define a set of insurable loans, then the constrain can be specified as:

$$\sum_i \sum_j \sum_k \sum_m l^t_{ijkm} L^t_{ijkm} \leq L^{t*}_{j=1} \qquad (31)$$

where $L^{t*}_{j=1}$ is the maximum amount of insured credit that the bank can issue.

There may also be requirements that the bank guarantees the supply of credit to the smallest far- mers or an explicit decision to exclude certain group. In this case specific constraints may be imposed on loan size, let's say:

$$\sum_i \sum_j \sum_k \sum_m 1^t_{ijkm} L^t_{ijkm} \geq L^{t*}_{k=1} \quad \text{or} \quad (32.a)$$

$$\sum_i \sum_j \sum_k \sum_m 1^t_{ijkm} L^t_{ijkm} \leq L^{t**}_{k=3} \quad (32.b)$$

The most severe institutional constraint under which agricultural development banks operate is the low nominal interest rate. Modeling this constraint does not require a specific equation, since it is embeded in the definition of the loans, i.e., in the column vectors of the model. Changing this parameter is an important policy consideration and some time is devoted to it in the empirical analysis undertaken in the following chapters.

An important set of constraints in the model are the physical resource constraints that, in the short run, define the bank's operating capacity. These constraints have been omitted from most bank port- folio models, under the assumption that, if a finan- cial decision proves rewarding, the institution will immediately adjust its resource base. In developing economies and particularly in public institutions, such adjustment is rather difficult to materialize. In the particular case of the ADBs, they have a limited number of loan officers, office capacity and equipment and vehicles for field supervision. On the other hand, previous research and bank officers' opinions support the hypothesis that loan supervision improves loan recovery rates. However, with current resources, ADBs are operating under severe pressure to administer many more loans than they should.

The specification of these constraints deserves further discussion. The situation can be modelled defining current loan requirements of vehicles and personnel (the stronger constraints) and current resources availabilities. Such relationships auto- matically define the current loan recovery rates. The constraints can be specified as follows:[11]

$$\text{personnel} \quad \sum_i \sum_j \sum_k \sum_m P^t_{ijkm} 1^t_{ijkm} \leq P^t \quad (33)$$

$$\text{vehicles} \quad \sum_i \sum_j \sum_k \sum_m v^t_{ijkm} 1^t_{ijkm} \leq B^t \quad (34)$$

where p_{ijkm} and v_{ijkm} stand for the personnel and vehicles requirements for the administration of each loan; and P^t and V^t are the resource supplies.

The set of loans could in this case be divided into loans with alternative levels of increasing supervision (w), let's say p_i^1; p_i^2... p_i^w; v_i^1, v_i^2 ... v_i^w, with their corresponding higher recovery rates. Specifying the above would imply knowing a functional relation between loan supervision and recovery rates. Hence, if the p_i^w and v_i^w are points of a linear relation of recovery rates and loan supervision, the constraints can be specified as:

$$\sum_w \sum_i p_{ijkm}^{wt} \, l_{ijkm}^{wt} \le P^t \tag{35}$$

$$\sum_w \sum_i v_{ijkm}^{wt} \, l_{ijkm}^{wt} \le V^t \tag{36}$$

Enlarging the availability of resources would have the following two effects: First, it would increase the total number of loans that can be issued; and second, it would increase the number of more supervision-intensive (more rewarding) loans.

Another important contraint on agricultural development bank refers to the usual decision on maximum allocation of funds to each regional office in each time period. This is done to allow each regional office to meet its credit obligations regardless of its performance in the previous period.
Hence, this constraint guarantees that poorly performing offices can continue in operation, while at the same time it deprives the best performing offices from having a faster growth. Once again in this case, the bank policy is to redistribute the benefits of credit among groups, benefiting those that because of "whatever reasons" were unable to pay their loans on time. Ultimately, such policy has resulted in severe indebtedness of poorly performing farmers who were given credit beyond their repayment capacity, while at the same time it has served to subsidize farmers that were unwilling to pay back their loans. This model does not include spatial disaggregation, hence this constraint is not included.

The institutional constraints defined above are the most important ones. The list can be expanded for particular situations without difficulty.

Financial Constraints

As any financial institution, but with certain peculiarities, ADBs face a series of financial constraints. The Capital Adequacy Ratio is perhaps one of the more stringent constraints in bank portfolio

and it is intended to preserve the banks as a viable
entity through periods of financial distress. Leverage requirements are, therefore, demanded by Central
Bank Authorities, as a balance between assets and
liabilities, depending on the liquidity characteristics of each component of the portfolio. It is not
known whether development banks are or not exposed to
penalties for not meeting these requirements, nevertheless good banking practices would demand them.

Dealing with assets, Chambers and Charnes (1961)
and Beazer (1975) suggest following the criteria of
the Examiners of the U.S. Federal Reserve Board.
Such criteria distinguishes four categories of assets,
each with a given illiquidity index (δ):

Category	Assets	δ
Primary and secondary reserve	Reserves and cash in vault and government securities maturing in less than two years	0.005
Minimum risk assets	Government securities maturing in more than two years and less than ten years and insured loans.	0.040
Intermediate assets	Other securities and government securities of more than ten years	0.060
Portfolio assets	Loans	0.100

It is possible on this basis to define a weighted
index of illiquidity of bank assets (A).

$$A = \sum_{i}^{I} \delta_i A_i$$

or more specifically:

$$A = 0.005 \, C + 0.040 \sum_i G_i + 0.060 \sum_i O_i + 0.100 \sum_i L_i \quad (38)$$

where all variables could be of different types:

On the liability side, the liquidity characteristics of the various sources of funds have also been
defined by the Examiners criteria.

Liability	η
Demand deposits	0.47
Time deposits	0.36
Other deposits and borrowings from Central Bank and from other banks	1.00

Therefore, one can also define a weighted index
of liabiliby liquidity (D):

$$D = 0.47 \ DD + 0.36 \ TD + 1.00 \ (BC + BI + GS) \qquad (39)$$

where all terms have been already defined.

On this basis, the leverage requirements can be expressed as:

$$A \leqq W - D \qquad (40)$$

where W is capital or net worth. Since each component of A and D is endogenous, we can express the leverage requirement as:

$$A + D \leqq W \qquad (41)$$

This constraint would encourage the bank to hold more liquid assets and less liquid liabilities. As it shifts more illiquid to liquid assets, the index A is reduced for a given level of total assets and as it shifts into less liquid liabilities, the index D is reduced for a given level of total liabilities. In either case, it becomes possible to increase total assets and total liabilities relative to net worth.

The banks are also recommended to hold a minimum amount of money determined as 15 percent of the first $100,000 of the portfolio, plus 10 percent of the next $100,000 plus 5 percent of the next $300,000. In the case for example of a bank with $500,000 in assets, the capital-adequacy ratio would be written as:

$$A + D \leqq W + 40,000^{12} \qquad (42)$$

The most important determinant of bank growth, is its returns from the activities of the previous period. As it was discussed in Chapter 2, the volume of loans and purchase of other assets in period t , depends on the collection minus costs (the net surplus of funds) in period (t-1), plus any authorized disbursements from borrowings from other banks, central government allowances, disbursable (net of reserves) time and demand deposits and operating and financial costs:

$$\sum_i \sum_j \sum_k \sum_m l^t_{ijkm} \ L^t_{ijkm} + \sum_i g^t_i \ G^t_i + C^t + [\ (1+Z).BC]^t + (OE)^t$$

| issued loans | + purchases of secu- | +cash+ | amortization of borrowed funds | + operating expenses |

$$\leqq [\Sigma \ l^{t-j,t}_{ijkm} \ L_{ijkm} \ (1 + r_{ijkm})\gamma_{ijkm}] + \Sigma \ g^{t-j}_i G^{t-j}_i \ (1+y_i)$$

(loan collections: principal and)   + maturing securities
                        interests

$$+ \quad (BC)^t \quad + \quad (BI)^t \quad + \quad (GS)^t \quad + \quad (TD)^t \quad + \quad (DD)^t$$

+ borrowings   + borrowings   + government   + time   + demand
   from   from IFAS   subsidies   deposits   deposits
   commercial   (net)   (net)
   banks

$$(43)$$

These constraints are the most important factors determinant of resource allocation by the ADB, according to specific allocative criteria. The following section takes up the issue of modelling the bank's objective to allocate financial resources among alternative uses over time, within the limits of existing constraints.

THE MULTIPERIOD MODEL

The Bank's Objective

As it was discussed in Chapter 2 an ADB has many goals to meet. These have been incorporated through constraints that demand credit for particular groups of producers and crops, plus the condition of low interest rates. However, as it also was discussed in Chapter 2, the banks must operate under some behavioral rule, and utility maximization has been chosen as this criterion.

The bank allocative criteria is necessarily based on dynamic cash balance relations. At each time period the allocation of funds is limited by the availability of resources and the risk and maturity preferences among the feasible choices of assets and liabilities. This determines that the bank maximizes a multiperiod financial objective, but meeting annual goals.

As for the objective of the bank, no generalization can be made as to what is the common rule in a dynamic decision process. There are a number of possible maximizable objectives, like net worth of the bank, total cash flow, discounted present value of cash flow, average earnings over some period, growth rate of earnings, utility measured over a period of time, etc. Although there are many possibilities, two have been used in most cases: a) maximize the net worth of the bank at the end of the planning horizon, and b) maximize the present value of the stream of cash flows generated during the planning horizon, plus some adjustment for increased net worth of the bank.

Bradley and Crane (1975) argue in favor of the first objective function because, they say, if all investment opportunities have been included in the model, it can be considered as a closed economy, completely describing the bank's earning opportunities. However, as the opportunity cost of money over time should be considered, particularly under inflation, a discount function may be preferred.[13]

Given the above, we can assume that the bank maximizes a discounted objective function, like equation (4) that now can be written as:

$$U = \sum_t [\frac{(NR)^t}{(1+\zeta)^t}] + \frac{TW_T}{(1+\zeta)^T} - \phi\Sigma \frac{\sigma_t}{(1+\zeta)^t} \qquad (44)$$

where $(NR)^t$ is net return, TW_T is terminal wealth at the end of the planning horizon, and everything else is already defined.

Notice that in the decision-making process what is important is to know the optimal decisions in the first period since these are the only ones that can be implemented. However, these "optimal first period decisions" must be interpreted carefully as the first step in a plan that is optimal in a long run sense. Later time periods are introduced because we do not know how to modify the criterion function to reflect the consequences of present decisions on future opportunities and because of intertemporal linkages.

In the context of multiperiod planning, two points deserve further attention: i) the choice of the discount rate, and ii) the length of the planning horizon. There is not complete agreement on these issues, hence the available alternatives are reviewed before proposing one:

i) Some of the plausible alternatives for the proper discount rate are: a) the bank's cost of capital; b) the rate at which the stock market implicitly capitalizes net income in determining the market value of the bank's stock, and; c) the bank's own and subjective time rate of preference for net income. Most economists would agree that, on a theoretical basis, the bank's cost of capital should be used. Furthermore, the nature of ADBs as public institutions provides a supporting argument for this criterion. To test the relative importance of this parameter, the model should be solved for different values of the discount rate.

It must be recalled that a time preference criteria is included because we recognize that there is the choice between savings and consumption in the Fisherian sense. Therefore, the model would allocate resources between end of period stockholder income and reinvestment. Since ADBs can in principle be public or private, the distribution of earnings at the end of each period will make sense only in the later case.

ii) Much discussion has also been offered in regard to the length of the planning period in a multiperiod LP model. The value of T in a T-period model should be chosen, such that the set of first period decisions becomes insensitive to further increases in T. More complex criteria are offered for the cases of replacement problem models, such as those applied for livestock and tree crop planning models (see Boussard, 1971; Weintgardner, 1962; Miller, 1979; and Gunter and Bender, 1980).

The above criteria gets complete validity when the model's objective function includes the terminal value of wealth. In such case one can obtain, with only T period solutions, the same first period decisions that would be made if returns were discounted over a much larger horizon. It should be mentioned however, that an important criterion in deciding for an initial length of planning period is the maturity of the bank's assets and liabilities. If the bank issues livestock loans that mature within a 10 year period and that is the longest maturity, then that is the minimum length of planning period if livestock loans were issued only in the first period.

The model proposed here, uses as a criteria for choosing T the "no change in the first period solution when another period is added". The activities in the model have maturities for assets and liabilities much smaller than the value of T.

Incorporating Risk Measures in the Multiperiod Model

The particular nature of agricultural banking implies that loans are a risky activity. The riskiness of bank earnings on loans is measured by the recovery rate. This reflects the forgone earnings because of late repayment or no repayment at all. At a particular point in time (s) the bank expects to recover $L_{is} + r_{is}$ (see Chapter 2) but it collects only $(L_{is} + r_{is})\gamma_{is}$. Hence the lower the value of γ_{is} the smallest the net return to the bank.

From previous experience the bank can estimate the expected recovery $E(L_i + r_i)$ as:

$$E(L_i + r_i) = \sum_{s=1}^{S} \gamma_{is}(L_{is} + r_{is})/S \qquad (45)$$

where γ_{is} is a random variable and (s=1...S) is the length of time periods for which the information is available. Similarly the variance of return can be expressed as:

$$V(L_i + r_i) = \frac{\sum_{s=1}^{S} [(L_{is} + r_{is}) - E(L_i + r_i)]^2}{S-1} \qquad (46)$$

Incorporating measures of risk in static mathematical programming models has followed alternative approaches. Because most of the work has used the E-V (Markowitz) approach, there has been heavy reliance on quadratic programming (QP) models. But also because of the computational difficulty with QP, alternative methods have been developed to approximate QP solutions by means of linear programming. Some of these include the mean of total absolute deviations (MOTAD) as proposed by Hazell (1971); separable programming, suggested by Thomas, et al. (1972), which approximates the non-linear total variance constraint by piece-wise linear functions and the marginal risk constraint LP, first suggested by Chen and Baker (1974), which consist of a complex multi-stage procedure.

Another approach with a limited applicability is the one proposed by McCarl and Tice (1980) which diagonalizes the variance-covariance matrix of returns by applying the principal axis theorem. The advantage of this approach is that it reduces computational complexities in applying separable programming by decreasing the number of piece-wise linear approximations; however, it is limited to the cases where the variance-covariance matrix is positive (or negative) definite. More recently, Kim and Yanagida (1981) developed Direction Constraint Linear Programming (DC-LP), which also approximated the variance boundary constraint.

MOTAD has been the most widely used technique for incorporating risk measures into LP models. It is simple and particularly attractive in large models that would be too expensive to solve in a QP format. Using MOTAD, with the advantages of a linear programming formulation is not without its problems. When the population of expected returns for each activity is approximately normally distributed and when estimates of the variance and the mean absolute deviation are used solely on sample data, though both estimated Standard Deviation and Mean Absolute Deviations (m.a.d.) are unbiased, the relative efficiencies of the two estimators may differ. Hazell (1971) had pointed out that the sample m.a.d. is only 88 percent as efficient as the standard deviation in estimating the population standard deviation.

Using the MOTAD approach demands that the variance of the loan portfolio, defined in equation (46) can be replaced by:

$$\sigma^2 = \sum_i \sum_s l_i \, l_{i'} [\frac{1}{S-1} (\dot{L}_{si} - \bar{L}_i)(\dot{L}_{si} - \bar{\bar{L}}_{i'})] \qquad (47)$$

where:

    s , is the number of periods for which sample observations are available (s = 1, 2, ...S)

    $\tilde{L}$, is the return (interest plus principal) to the ith loan in period s. Notice that i' is used to denote any other loan.

    $\bar{L}$ is the mean return for the ith loan, which is defined

$$\bar{L} = \frac{\sum_{s} L_{si}}{S} \tag{48}$$

Summing up over s in equation (47) and factoring:

$$V = \frac{1}{S-1} \sum_{s=1}^{S} [\sum_{i=1}^{n} L_{si} l_i - \sum_{i=1}^{n} \tilde{L}_i l_i]^2 \tag{49}$$

where $\sum_{i=1}^{n} \tilde{L}_{si} l_i$, is the total gross return of a particular loan portfolio.

Assuming that the same sample data are available as for equation (49), the M.A.D. (denoted by Γ), may be defined as:

$$\Gamma = \frac{1}{S} \sum_{s=1}^{S} \left| \sum_{i=1}^{n} (\tilde{L}_{si} - \bar{L}_i) l_i \right| \tag{50}$$

which is an unbiased estimator of the population m.a.d. Hence, the (E,V) model can be replaced as the (E,Γ) model. The advantage of the E,Γ model is that it leads to an LP formulation of the portfolio selection problem. When returns exhibit a trend, the deviations can be measured from the regression line of returns over time as shown by Hazell and Pomareda (1981).

To approximate Γ as an estimator of the standard deviation, it is possible to write:

$$\tilde{\sigma}_i = \frac{\Delta^{\frac{1}{2}}}{S} \sum_{s=1}^{S} (\tilde{L}_{si} - \bar{L}_i) l_i \tag{51}$$

where $\Delta^{\frac{1}{2}}$ is a correction factor to convert the mean absolute deviation to an estimate of the population standard deviation, as proposed by Hazell and Scandizzo (1974).[14] This formulation can be entered in the model by defining new variables Z ⪈ 0, and forming the problem:

$$\max U = \sum_{i=1} E(\tilde{L}_i) - \phi\sigma \tag{52}$$

subject to:

$$\Sigma(\bar{L}_{si} - \bar{L}_i)1_i + Z_s \geq 0, \quad s = 1,\ldots S \qquad (53)$$

and

$$\sum_{s=1}^{S} Z_s - \bar{\sigma} \frac{S}{2} \frac{1}{\Delta^2} = 0 \qquad (54)$$

The Z variables measure the negative deviations of returns from the mean; therefore $2 \sum_s Z_s$ is the sum of total deviations. In matrix form, for a one year period, the problem can be formulated as in Figure 3.3.

This formulation is in general applicable in a static model. However, as it was discussed earlier, the multiperiod objective function of the bank equation (44) implies a measure of covariance of returns between time periods. We should recall that the LP model uses an estimate of the standard deviation. Therefore, assuming only two time periods, equation (28.a) can be written as:

$$V(R) = V(R)_1 + V(R)_2 + 2\rho_{1,2}\sigma(R)_1\ \sigma(R)_2 \qquad (55)$$

which means that only when the correlation coefficient ($\rho_{t,t+1}$) is equal to 1, the standard deviation is:

$$\sigma(R) = (R)_1 + (R)_2 \qquad (56)$$

which, as it would be expected, simplifies considerably the Linear Programming formulation of intertemporal risk linkages. If $\rho_{t,t+1}$ is not equal to one, a different formulation is needed.

The standard deviations of returns in each period are calculated through the MOTAD approximation procedure. The correlation coefficient can be estimated from bank data for various years.

The Model in Tabular Form for an n-Period Horizon

There are three basic types of intertemporal linkages in the multiperiod model: linkages in the asset portfolio, linkages in the liability portfolio and linkages in the risk accounts.

Intertemporal linkages in assets exists in the loan and investment portfolios. There are loans and bonds with maturity of more than one period, therefore; if issued in year t-j , they are collected in year t, but also they use physical resources during

Columns / Rows	L O A N S				MAD ACCOUNTS				RISK AV. COEF.	RHS
	l_1	l_2	$\cdots$	l_n	z_1	$z_2 \cdots$		z_n		
U	$\bar{L}_1$	$\bar{L}_2$	$\cdots$	$\bar{L}_n$					$-\phi$	$=$ MAX
z_1	$\bar{L}_{11}-\bar{L}_1$	$\bar{L}_{12}-\bar{L}_2$	$\cdots$	$\bar{L}_{1n}-\bar{L}_n$	1					≥ 0
z_2	$\bar{L}_{21}-\bar{L}_1$	$\bar{L}_{22}-\bar{L}_2$	$\cdots$	$\bar{L}_{2n}-\bar{L}_n$		1				≥ 0
$\vdots$							$\cdot$			$\cdot$
z_s	$\bar{L}_{s1}-\bar{L}_1$	$\bar{L}_{s2}-\bar{L}_2$	$\cdots$	$\bar{L}_{sn}-\bar{L}_n$				1		≥ 0
$2\sum_{s=1}^{T} z_s$					2	2 $\cdots$		2	$-\left[\dfrac{1}{\Delta^2\,S}\right]$	$= 0$

(Rows $z_1 \ldots z_s$ are labeled MAD CONSTRAINTS.)

Figure 3.3   The MOTAD Formulation for One Planning Period.

Table 3.1
Description and Code of Variables

Nominal Value	Description	Number

Assets

L	Loans (principal)	l
G	Government securities	g
C	Cash	-

Liabilities

DD	Demand deposits	dd
TD	Time deposits	td
BC	Borrowings from CB's	-
BI	Borrowings from IFA's	-
GS	Government subsidies	-

Subscripts

i	Purpose of loan
j	Insured (=1) or not insured (=0)
k	Size (any asset or liability k=1---K)
n	Maturity (any asset)

Superscript

t	time periods

Other

δ	Illiquidity index of assets
η	Liquidity index of liabilities
A	Weighted index of illiquidity of bank assets
D	Weighted index of liquidity of bank liabilities
γ	Loan recovery rate When it appears our capital letter describing an asset it indicates the expected collection at maturity (i.e. L)
I	Return on securities

Table 3.1 (continued)

Nominal Value	Description	Number
Other		
r	Rate of interest on loans	
z	Rate of interest on borrowed funds	
Δ	Fisher's coefficient to approximate the mean absolute deviation in an estimator of the standard deviation	
π	3.1416	
S	Number of years of available information to calculate mean absolute deviations	
Γ	Estimator of the standard deviation	
σ	Standard deviation	
φ	Risk aversion coefficient	
ρ	Correlation coefficient	
λ	Discount factor	
ζ	Price of capital	
TW	Terminal wealth	
R	Return	
U	Utility	
W	Bank capital or net worth	
OE	Operating expenditures	
P	Personnel cost	
V	Vehicle maintenance and repairs	
O	Other expenditures	

their life. Similarly, in the liability side, the bank's borrowings are repaid after several periods, although there may be continuous payment of interest costs.

The intertemporal linkages establish that net receipts in year t-1 are fully transferred to year t, hence the model does not include an annual income disbursement activity. This is equivalent to saying that there is not a payment of end of period dividends, only an accumulation of wealth.

The formulated intertemporal risk linkages assume a perfect correlation of returns over time. The matrices of mean of absolute deviations for each period are discounted and added and taken into account in the objective function with a given weight. This weight is given by the risk aversion coefficient (ϕ). A value of ϕ=0 may reflect the attitude of a public institution, fully supported by the government, in which case the risk of returns is not important. Larger absolute values of ϕ will imply more concern with risk management. Assuming that returns are normally distributed, the value of ϕ can be interpreted as the level of confidence for a given standard deviation of returns. Following Baumol (1963) then values of ϕ=1.65 and ϕ=3.15 correspond to investors maximizing the 0.05 and 0.001 percentiles of their income distributions, provided these are normally distributed. Therefore, the values of ϕ = 1.65 and ϕ =3.15 are defined as 'reasonable' and 'extreme' levels of risk aversion, respectively.

NOTES

1. The appropriate definition of return can vary among investors.

2. Markowitz' original work was published in 1952 and further extended in 1959. The last edition of his book published in 1976 provides a most complete exposition and a very extensive review of literature.

3. For discussion purposes "investments" are used as a generalization, yet they can represent securities, loans, bonds, or any other type of investment activity.

4. Quoted by Savage (1954).

5. There are alternatives to the Markowitz-Tobin approach. Baumol (1963) offered the EL Criterion, expressed as a lower confidence limit on E equal to $L=E-K\sigma$ where E represents expected returns, K is a constant determining the probability with which E exceeds L and σ is the standard deviation of returns.

6. See Hazell (1971), and Pomareda and Simmons (1977).

7. In fact Beazer (1975, p. 13) interprets this model as a modification of the Markowitz model in the sense that constraints are applied to eliminate the necessity of considering risk in the objective function. But it should be recognized that this approach does not take into account covariance effects.

8. An interesting discussion of the concepts involved in determining the relevant planning horizon was provided by Modigliani and Cohen (1961). This issue is discussed further in Chapter 4.

9 In any LP model, the number of vectors in the solution cannot exceed the number of effective constraints.

10. Crops, Livestock, farm improvements and other.

11. The subindices t, j, k and m could be omitted for simplicity. That however, does imply that resource requirements do not change among the different types of loans and time periods.

12. (0.15) (100,00) + (0.10) (100,00) + (0.05) (300,000) = 40,000.

13. Cohen and Hammer (1966, 1967) used a discounted objective function; however they recognized that the choice of the discount rate definitely affects the optimal solution. Higher discount rates will emphasize generating cash flows earlier. Bradley and Crane (1975) used undiscounted objective functions.

14. The value of $\Delta^{\frac{1}{2}}$ is calculated as a function of the number of periods for which information is available (S) and $\pi = 3.1416$.

15. In each time period the model includes several assets and liabilities, but the general rule for specification of signs in the matrix is that all receipts of funds and supply of resources are denoted negative, while the uses of funds and demand for resources are denoted positive.

3
A Model for the Agricultural Development Bank of Panama (ADBP)

BACKGROUND

The Economy and the Agricultural Sector

Panama is one of the smallest countries in Latin America with a population of near two million people, and an annual percapita GNP of only US$660. Average per capital GNP has not increased during the past decade, yet the distribution of income has improved. This has been largely due to a significant amount of public investment in social programs. The latter has put the country in serious indebtness, as the public debt rose from US$237.75 million in 1969 to US$1645.26 million in 1979. Furthermore, the public debt in 1979 represented 85.5 percent of GNP, while in 1970 it accounted for only 24.7 percent of GNP (IDB, 1981; p. 107) as shown in Table 4.1.

As in other Latin American countries, the agricultural sector is of major importancce in the local economy. Nevertheless, the contribution of agriculture to GNP has decreased from 19.4 percent in 1969 to 16.0 percent in 1978. The sector of commerce and international trade is the most important, and this is to a great extent due to the existence of the Panama Canal.

The country's main exports are coffee, bananas, shrimp and oil derivatives, but they have not increased enough to reduce the balance of payments deficit. This has widened considerably in the last decade, as imports of oil and capital inputs rose steadily. Also food imports are an important component in the overall import bill, accounting for approximately 7 percent. In absolute terms, food imports rose from US$20.85 million in 1969 to US$61.03 million in 1978.

Panama's tax laws have encouraged international commercial banking, hence Panama City has become the banking center for Latin America. There are

71

Table 4.1
Panama, Major Economic Indicators, 1971-1978

Variable	Unit	1971	1972	1973	1974	1975	1976	1977	1978
Gross national product (In 1960 prices)	Mill.US$	973	1034	1101	1130	1137	1133	1172	1204
Agric. as percent of GNP	%	17.2	16.6	16.0	15.5	16.1	16.1	16.8	16.0
Per capita GNP In 1960 prices)	Mill. US$	648	670	694	693	678	657	660	660
Government debt	Mill. US$	328	405	476	590	772.	855	956	1412
Number of bank offices		140	152	169	178	185	187	184	181
Capital assets of all banks	Mill. US$	1167	1888	3578	6475	8433	9865	12435	16064
Outstanding domestic loans	Mill. US$	563	764	1025	1352	1525	1658	1765	1843
Outstanding foreign loans	Mill. US$	375	608	1208	2891	4461	4743	6280	8495

Food imports	Mill. US$	34	33	42	55	55	54	58	61
Total imports	Mill. US$	359	401	454	755	815	779	777	844
Total exports	Mill. US$	114	121	135	204	280	228	243	244
Consumer price index (1975=100)		72	75	81	94	100	104	108	113
Index of agric. prices (1975=100)		100	102	107	126	142	146	157	175

Source: Panamá, Contraloría General de la República, Panamá en Cifras, Contraloría General de la República, Panamá, 1979.

currently 115 foreign and domestic banks operating in
Panama with a total of 181 offices. The importance
of this activity is significant in the local economy,
as it provides employment for several thousand people,
yet most of the financial resources are channeled
towards other Latin American countries. In 1969 the
outstanding loans to domestic borrowers were
US$331.81 million, while those to foreign borrowers
were only US$101.89 million. By 1979 the proportions
reversed dramatically and outstanding domestic loans
were US$1,843.25 million while foreign loans were
US$8,495.22 million.

The flourishing of the international banking
industry has attracted some of the best local talent.
As a result the two national development banks face
serious management and personnel problems as they
cannot offer competitive salaries and privileges.
Bernal, Herrera and Joly (1982) suggest that this
lack of personnel is one of the major reasons for a
defficient service and low loan repayment in the Agricultural Development Bank of Panama (ADBP).

The agricultural sector of Panama is tipified by
small farms. As shown in Table 4.2, more than 95
percent of the farms are of less than 50 hectares,
while only a few farms are of more than 500 hectares.
The latter group includes sugar cane plantations and
some large rice producing companies.

As shown in table 4.3 rice and corn are the two
most important food crops, while coffee, tobacco and
sugar cane are the major export crops. As evidenced
in the analysis of the following sections, formal
government and private credit is assigned mostly to
these crops and cattle.

Financial Resources and Policies for Agriculture

As in most developing countries there is a strat-
ification in the agricultural credit market. Commercial banks and the two development banks are the
source of institutional credit, each serving a particular clientele and operating under different
criterion.

Commercial banks supply the largest proportion of
credit to agriculture, but practically all of it is
allocated to large enterprises producing sugar cane,
rice and beef cattle (see Table 4.5). Most of these
loans are very large and in fact the Fiduciary Bank
and the Chase Manhattan Bank do not provide loans of
less than US$25,000 and US$15,000 respectively
(Bernal, Herrera and Joly, 1982). In the selection
of their clientele, commercial banks rely on past
records and analysis of profitability of the enterprises as the most important criteria.

Table 4.2
Panama, Land Use and Size Distribution of Farms

Variable	1950	1960	1970
Land Use			
Number of farms	85,473	95,505	105,272
Total Acreage (has)[a]	1.159,082	1,806,452	2,098,062
Annual crops	–	205,048	213,607
Perennial crops	–	125,378	110,764
Cultivated pastures	427,557	683,606	964,758
Natural pastures	124,530	134,723	176,037
Fallow lands	213,563	222,971	217,436
Other lands	156,820	134,726	415,460
Size Distribution of Farms (has)			
< 50	81,755	88,713	96,654
50-100	2,407	4,329	5,526
100-500	1,157	2,239	2,773
>500	154	224	319

Source: Panamá, Contraloría General de la República, Panamá en Cifras, 1974-1978; Panama, 1979.

[a]One hectare (ha) = 2.45 acres.

Table 4.3
Panama, Area and Production of Main Crops

Area of Main Food Crops (has)	1974/75	1975/76	1976/77	1977/78	1978/79
Rice	112,200	115,370	122,350	109,980	99,110
Corn	75,500	74,320	83,150	82,780	68,600
Beans	16,100	16,590	15,560	14.850	11,770
Production of Main Export Crops					
Coffee (1000qq)[a]	99,300	105,850	103,100	121,600	135,500
Tobacco (1000qq)[a]	17,660	24,510	28,310	30,650	33,290
Sugar cane	1,898	2,121	2,,641	3,039	2,892

Source: Panama, Contraloría General de la República, Panamá en Cifras, 1974-1978. Panama, 1979.

[a] 1 quintal (qq) = 46 kg.

Table 4.4

Panama, Credit from Public and Private Sources by Sector (million US$)

Activities	1975	1976	1977	1978	1979[a]
Public	266.21	351.02	334.82	209.56	396.40
Domestic Sector	260.21	351.02	334.83	209.56	376.40
Agriculture	5.68	3.25	7.75	4.32	7.61
Livestock	17.35	14.74	28.11	22.74	25.63
Fisheries	0.01	0.01	0.02	-	-
Other activities	237.17	333.02	298.96	182.30	343.56
External Sector	6.0	-	-	0.20	-
Private	8,579.93	10,147.14	9,905.32	14,469.20	19,564.21
Domestic Sector	1,647.39	1,631.14	1,664.54	1,853.25	2,282.68
Agriculture	69.28	69.49	88.87	66.85	68.39
Livestock	48.45	39.43	36.23	39.77	40.93
Fisheries	8.55	4.54	7.08	5.51	4.17
Other activities	1,521.11	1,517.68	1,532.36	1,741.12	2,169.19
External Sector	6,932.53	8,516.01	8,240.78	12,615.78	17,281.53
Total	8,846.14	10,498.16	10,240.14	14,678.76	19,940.61

Source: Panamá, Comisión Bancaria Nacional, 1981.

[a]Preliminary

Table 4.5
Panama, Credit from Private Banks[a] to the
Agricultural Sector (1,000 US$)

Purpose	1977	1978	1979	1980
Crops				
Rice	28,860	22,698	23,334	25,443
Sugar cane	53,260	33,996	32,908	36,219
Coffee	3,591	6,581	8,648	5,057
Other crops	3,198	3,572	3,500	3,533
Total	88,869	66,847	68,390	70,254
Livestock				
Cattle	29,388	33,380	32,522	35,866
Chicken	5,365	4,075	6,399	5,431
Other	1,574	2,217	2,006	3,221
Total	36,227	39,772	40,927	44,418
Total	125,096	106,619	109,317	114,672

Source: Panamá, Dirección de Estadística y Censo y
Comisión Bancaria.

[a]Includes City Bank, Chase Manhattan Bank, Banco
Fiduciario, Banco de Colombia, Banco de Santander,
Bank of America, Banco de Comercio Exterior, Marine
Midland Bank, Sociedad de Bancos Suizos de Panamá,
and Banco Internacional.

Table 4.6
Panama, Agricultural and Other Credit from the National Development Bank, BNP (1,000 US$)

Sector	1976-77	1977-78	1978-79	1979-80
Agriculture	7,745	4,320	7,206	7,250
Livestock	28,104	22,748	26,114	30,516
Other sectors	266,709	147,038	347,670	458,617
Total	302,559	174,102	381,010	516,389

Source: Panamá, Banco Nacional de Panamá, Memoria Anual, various issues.

Table 4.7
Panama, Crop and Livestock Credit Provided by the Agricultural
Development Bank (1000 US$)

Purpose	1977-78	1978-79	1979-80	1980-81
Rice	7,177	7,632	12,962	13,522
Coffee & cocoa	-	722	3,524	3,423
Vegetables[a]	1,053	1,399	2,047	2,959
Corn & sorghum	2,477	2,432	3,457	4,622
Industrial tomatoes	1,038	1,068	1,063	1,806
Cattle	6,261	8,254	11,249	15,446
Other livestock[b]	668	1,619	2,485	1,854
Other loans[c]	1,021	1,664	2,575	3,072
				-
Total Credit	19,695	24,790	39,362	47,704

Source: Panama, Banco de Desarrollo Agropecuario. Memoria
Anual, various issues.

[a]Includes onions, potatoes, melons, and other vegetables.
[b]Includes poultry, pigs, and minor species.
[c]Includes beans, sugar cane, and other crops.

Table 4.8
ISA—Summary of Insurance Operations, 1976/77 – 1980/81

Variable	1976/77	1977/78	1978/79	1979/80	1980/81
Combined Portfolio					
Coverage (US$)	25,989	1,129,579	2,636,498	8,131,592	13,114,208
Number of policies issued	9	351	809	2,114	2,772
Indemnities paid (US$)	1,588	17,784	102,462	194,642	402,143
Net premiums (US$)	1,165	58,723	113,815	331,567	519,579
Loss ratio	1.36	0.30	0.90	0.59	0.77
Crop Insurance					
Coverage (US$)	25,898	1,129,579	1,887,511	4,575,710	6,806,637
Hectares insured	122	5,410	7,307	13,988	16,183
Number of policies issued	9	351	525	1,284	1,446
Indemnities paid (US$)	1,588	17,784	93,731	130,451	290,013
Net premiums (US$)	1,165	58,723	103,741	269,630	356,261
Loss ratio	1.36	0.30	0.90	0.48	0.81
Livestock Insurance					
Coverage (US$)			748,987	3,555,882	6,307,571
Number of head insured			3,392	11,677	18,969
Number of policies issued			284	830	1,276
Indemnities paid (US$)			8,731	64,191	112,130
Net premiums (US$)			10,074	61,937	163,319
Loss ratio			0.87	1.04	0.69

Source: ISA, Memoria Anual, various issues.

The National Development Bank (BNP) is one of the two government banks. As shown in Table 4.6 it serves primarily other sectors, but approximately 10 percent of its portfolio is allocated to agricultural and livestock loans of medium and large size. Inspite of being a development bank, it follows quite strong loan selection criteria and it is tipified by a rather high loan recovery rate.

The Agricultural Development Bank of Panama (ADBP), is the government bank that serves exclusively the agricultural sector. It provides credit for small, medium and large individual farmers and cooperative groups. Its portfolio is widely diversified by crops and livestock in the nine provinces of Panama. The bank provides credit at the lowest interest rates and relies heavily on government subsidies. More on the structure of the ADBP will be provided in the following section.

Credit for agriculture provided by all banks is under preferential conditions. In October of 1980 the National Banking Commission set the interest rate on agricultural loans by commercial banks equal to 14.25 percent, while the market rate was 18 percent. Under these conditions, clients were granted a subsidy that between October 1980 and October 1981 accounted for US$4.5 million (La Estrella de Panamâ, January 15, 1982). Prior to 1980 the National Bank Commission,, pursuant to article 1, Law 95 of November 1974, had been granting a similar compensation to banks and finance institutions that provided credit to agriculture and manufacturing industries. The BNP supplies credit at rates between 12 and 14.5 percent and the loans from the ADBP vary between 8 to 11 percent.

The Agricultural Credit Insurance Program

The Agricultural Insurance Institute (ISA) began its operations in 1976 in close liasson with the ADBP. The program provides credit insurance and hence in first instance it protects the loans issued by the ADBP. The program has expanded rapidly (see Table 4.8) and currently insurers approximately 30 percent of the ADBP portfolio.[1] It insurers only rice, corn, sorghum, beans, industrial tomatoes, and beef cattle. It operates only in 6 of Panama's 9 provinces. The administration of ISA plans to cover an increased proportion of the ADBP portfolio, yet this would depend on the rate of growth of the ADBP activities and the desirability for insurance among self financed farmers and commercial banks.

Under the current rules and regulations of ISA, the total coverage level for a crop is 70 percent of the sum of the direct production costs. However, when a compensation payment is made, actual production costs are used when these are smaller than the stipulated production costs in the program. Somewhat asymetrically, when an actual production cost is higher than it was projected to be in the crop program, the lower figure is always used. Furthermore, compensation payments are made on the basis of the costs incurred up to the date of the disaster. Since a large proportion of the crop losses come soon after planting, this means that the actual coverage level for most cases where compensation payments are made is only about 50-60 percent of the total production costs. This has been a sore point, and the ADBP recently used its infuence to get ISA to pay farmers experiencing complete loss of crop the full value of the insurance coverage. Hogan (1981) suggested that this is going to have a significant impact on ISA's loss ratios. In this, ISA is facing a situation similar to that faced the Mexican insurer, (ANAGSA) in the SAM pilot areas.[2] In Mexico, insurance is being used as a contingent income transfer, in that the compensation payment exceeds the actual coverage (Hogan, 1981).

The administrative costs of ISA run about 6% per dollar of coverage, which makes it slightly larger than the average premium rate (ISA, 1980). ISA's average loss ratio for the last three years is approximately 73% for its crop portfolio and roughly 86% for its livestock portfolio. ISA receives administrative subsidies from the Government of Panama that account for approximately 2 percent of total coverage.

Within the agricultural sector two other institutions play an important role in the effectiveness of credit policies and the use of agricultural credit. The Ministry of Agricultural and Livestock Development (MIDA) is responsible for supplying the ABP and ISA the technical information for the crop and livestock programs which will receive official credit and insurance. The MIDA takes a somewhat more predominant role in overall agricultural sector credit planning, as it is also responsible for providing the technical assistance to credit users, and it has the responsibility for the creation and continuing support of the agrarian reform settlements.

The Agricultural and Livestock Marketing Institute (IMA) of Panama sets price guarantees each year which are used in planning the next crop production cycle. These prices are used in establishing loan and insurance programs. IMA also functions as a crop storage agency, although there are frequent reports of storage capacity being exceeded, with the result that farmers are often forced to sell to local traders at prices below the guarantee. Thus, price

stabilization has a rather qualified meaning in
Panama. Lastly, IMA is responsible for importing
foodstuffs which local producers are unable to supply
in sufficient quantity--principally maize, beans, and
sometimes rice. Again, price stabilization and food
import policies are normally coordinated with technology, credit, and insurance policies, but an ade-
quate understanding of how these policies interrelate
is still lacking.

THE ADBP PORTFOLIO

Financial Structure of the ADBP

The ADBP is the government bank to serve the
agricultural sector. As such, it fulfills government
goals and policies; yet its current institutional
design, availability and quality of resources, and
financial policies limits its capacity to provide
enough financial resources to agriculture. Furthermore, unless significant changes are introduced, it
is not likely that the bank could grow at a faster
rate. Nevertheless, the current administration is
very interested in exploring alternative growth
strategies.

The ADBP's balance sheet (Table 4.9) is typical
of the several specialized state-owned agricultural
development banks. On the asset side, the bank holds
basically loans and only in the last two years has it
included some bonds of government debt. On the
liability side, the bank used to depend primarily on
long term borrowed funds, yet short term borrowings
have recently increased considerably. The bank's net
capital accounts for most of total capital, but as it
is observed, the government contribution is also a
major portion of total capital. In significant
contrast with a liability-diversified bank, the ADBP
does not handle any savings and checking accounts.

The importance of loan recovery on the sources of
funds is appreciated from the data in Table 4.10.
Loan recovery represents 72 percent of internal
resources. Interest earnings on loans, on the other
hand, account for only 11 percent of total internal
resources. As borrowing from international agencies
has become more difficult, the bank relies currently
on heavy borrowings from commercial banks. These
borrowings currently account for 73 percent of external resources. Furthermore, an increasing portion
of these funds are short term borrowings.

The uses of funds reveal the high operating costs,
and strong demands for repayment of borrowed funds.
Hence, the bank is left with slightly over fifty percent of its resources available for loans. It is

Table 4.9
ADEP, Balance Sheet, 1979–80 (million US$)

Assets			Liabilities & Capital		
Description	1978/79	1979/80	Description	1978/79	1979/80
Cash	1,754	1,698	Borrowed funds:		
			Short term	3,843	9,110
			Long term	17,416	22,684
Loan outstanding	60,333	60,796	Pending accounts & bonds:		
			Short term	2,366	2,906
			Long term	1,886	1,886
Accounts outstanding	8,289	12,712	Other liabilities	0,541	0,550
Investments (nego-tiable)	0,054	8,647a	Interest earnings not collected	2,049	–
Real assets	1,517	1,590	Total liabilities	28,113	37,128
Fixed assets	7,828	1,590	Net capital	37,268	37,268
Other assets	0,700	0,602	Surplus (deficit)	2,082	0,448
			Contribution of the government	17,955	20,811
			Net return (Loss)	0,874	0,938
			Total capital	52,265	57,589
Total assets	80,379	94,717	Total liabilities & capital	80,379	94,717

Source: Panamá, Banco de Desarrollo Agropecuario, Memoria Anual, 1980.

aIncludes Treasury Debt for US$8,600 million.

Table 4.10
ADBP, Sources and Uses of Funds, 1979/80 and 1980/81 (million US$)

Sources of Funds	1979/80	1980/81
Internal Resources		
Loan recovery	23,809	25,365
Interest earnings	3,747	4,802
Government subsidy	2,943	6,119
Other resources[a]	2,369	1,535
Total	32,869	37,823
External Resources		
Borrowing from domestic banks	21,400	22,739
Borrowing from international agencies		
IDB	5,915	8,311
USAID	1,179	0,069
World Bank	0,780	0,676
Total	29,274	31,795
Total Sources	62,143	69,618

Uses of Funds	1979/80	1980/81
Operating Expenses		
Salaries & honorariums	3,327	3,572
Other operating expenses[b]	0,870	1,517
Capital disbursements	0,219	0,214
Total	4,416	5,304
Financial Expenses		
Repayment of borrowed funds	17,379	19,391
Interest payments	2,611	3,802
Other obligations	0,387	,099
Total	20,387	23,293
Loans	33,388[c]	37,869
Total Uses	62,143	69,618

[a] Sales of property and others.
[b] Includes vehicles, maintenance of offices, equipment, etc.
[c] This number is not equal to the one on Table 4.9 (39,362), hence here the ADBP includes only actually disbursed funds,, while the 39,362 includes committed but not fully disbursed funds.

evident that issuance of new loans is severely
affected by current financial obligations and the
operating budget of the bank

Characteristics of the Loan Portfolio

Lending is the primary activity of the bank, and
therefore; an examination of the characteristics of
the loan portfolio became an essential part of this
research. The bank did not have processed information on the characteristics of its loan portfolio
disaggregated by crop, size of operation, insurance
usage and maturity, as needed for the purpose of this
research. Also the available aggregated information
by crops and livestock loans did not provide data on
the characteristics of each loan in terms of amount
disbursed, interest rate, amount collected, administration costs, maturity structure of amortization,
etc.

To obtain the required information, consistent
with the model proposed in chapter 3, a sample of
loans issued by the eight most important agencies was
taken. A completely random sample was drawn from the
agencies files for the years 1974 through 1980
seeking approximately 10 percent of the loans issued
in each agency in each year, for each of the seven
major items shown in Table 4.7. The original sample
included 1,366 loans.

It was realized, during the processing of the
loans, that most of the medium and long term loans
issued after 1977/78 had not matured yet, hence making this set useless for the analysis. Those observations were dropped from the file, leaving a total
of 900 observations.

The set was desaggregated by size class and insurance class. The size classes were in U.S. dollars:

$$k = 1, \qquad < 10,000$$
$$k = 2, \quad 1,000 - 10,000$$
$$k = 3, \qquad > 10,000$$

and the insurance classes included:

$$\text{insured } (j = 1) \text{ and}$$
$$\text{not insured } (j = 0)$$

Hence, for each item there could be as many as
six classes. Therefore, there could be as many as 42
loan groups (7x3x2). In practice however, the number
of classes is a smaller set. This is because corn,
coffee and industrial tomatoes are produced only by
small farmers. Although rice and livestock are
produced by all groups of farmers; the insurance

Table 4.11
ADBP, Characteristics of the Not Insured Loans

Product	Size	Amount Disbursed ($)	Nominal Rate of Interest (%)	Amount Collected ($)	Net Interest ($)	Actual Rate of Interest (%)	Expected Duration (months)	Actual Duration (months)
Rice	1	449	9.25	475	27	6.41	7.85	13.20
	2	5,013	9.61	5,257	235	5.33	7.60	11.74
	3	21,638	9.69	22,520	882	5.36	8.43	12.06
Corn	1	440	8.88	473	33	7.57	8.57	12.81
	2	1,366	8.84	1,473	106	6.73	8.38	13.50
Industrial	1	433	8.69	457	36	5.64	4.55	14.17
Tomatoes	2	1,610	9.00	1,704	86	7.08	5.91	12.80
Vegetables	1	505	8.75	530	24	5.72	6.22	11.70
	2	2,376	9.37	2,500	153	6.80	6.00	9.96
Coffee	1	490	8.59	524	33	6.46	10.81	12.58
	2	2,447	9.13	2,594	147	6.40	10.51	12.51
Livestock	1	611	9.17	704	93	6.95	31.33	28.33
	2	3,988	9.60	4,574	586	6.73	37.36	33.24
Other loans	1	513	8.97	540	28	6.58	15.57	15.24
	2	2,456	9.26	2,635	179	6.60	11.36	13.39
Associated Producers	3	36,256	8.44	37,867	1.611	4.69	14.10	13.79

Source: Banco de Desarrollo Agropecuario de Panamá, Sample of 900 loans issued between 1974 and 1980.

Note: 1 = small loans of less than US$1,000
2 = medium loans of more than US$1,000 but less of US$10,000
3 = large loans of more than US$10,000

Table 4.12
ADBP, Characteristics of the Insured Loans

Product	Size	Amount Disbursed ($)	Nominal Rate of Interest (%)	Amount Collected ($)	Net Interest ($)	Actual Rate of Interest (%)	Expected Duration (months)	Actual Duration (months)
Rice	1	589	10.10	606	17	8.67	5.83	4.00
	2	4,722	10.36	4,886	164	5.86	7.69	7.19
	3	22,275	11.6	23,299	1,233	6.78	8.01	8.33
Corn	1	748	9.5	799	51	9.24	6.50	8.58
	2	2,286	9.96	2,387	101	8.75	6.79	8.29
Industrial	1	394	9.17	403	11.3	8.86	5.06	4.50
tomatoes	2	1,597	10.83	1,658	61	9.95	5.67	4,83
Vegetables	1	–	–	–	–	–	–	–
	2	–				–	–	–
Coffee	1	–	–	–	–	–	–	–
	2	–	–	–	–	–	–	–
Livestock	1	–	–	–	–	–	–	–
Other loans	2	5,034	10.08	5,633	599	9.42	23.26	15.08
	1	693	9.27	718	25	6.58	5.08	6.67
	2	3,703	10.07	3,878	174	6.49	5.24	8.85
Associative credit	3	–	–	–	–	–	–	–

Source: Banco de Desarrollo Agropecuario de Panamá, Sample of 900 loans issued between 1974 and 1980.

Note: 1 = small loans of less than US$1,000
2 = medium loans of more than US$1,000 but less of US$10,000
3 = large loans of more than US$10,000

institute does not insure the very large loans nor
the very small ones for individual farmers. Finally,
ISA insures only three of the five crops and
livestock.

An important part of the bank's loans are given
to cooperatives and farm associations. In 1979/80
and 1980/81 this form of credit accounted for 18 and
21 percent of the total volume of credit. Most of
this however, (approximately 65 percent) was for rice
production. This form of credit was provided in
loans of more than $20,000 and some of them were as
large as $200,000. Currently the program services
200 farmers' groups that included approximately 8,000
families; and 46 cooperatives with near 20,000
members.

The cooperative credit has resulted in a mixed
experience for the bank. On one hand, it reduces
considerably the administration costs. On the other
hand, the cooperative movement in the agricultural
sector of Panama is not very successful and this contributes to low repayment on bank loans. Given the
importance of this line of credit, and additional
loan category was included: Large loans for associated rice producers. Information was not available
through the sample of other items, nor for insured
rice for this group of producers.

As shown in Table 4.11 and 4.12, there were 16
classes of not insured loans and 10 classes of insured loans. The summary information for each loan
class revealed important characteristics. First,
there is a significant difference in the maturity of
crop loans versus livestock loans, as the later are
for the purpose of herd development, although some
may be for fattening. Second, for the non-insured
loans the actual maturity (average repayment period)
is significantly longer than the expected maturity.
For the insured loans, however, the actual maturity
is equal, and in some cases, even smaller than the
expected maturity. This implies that the administration cost of insured loans is smaller. Third, the
interest rate on loans for the same crop (except
rice) increases with the size of loan. The data
reveal also important differences in profitability of
the loans. In more than one case this profitability
was found to be negative, which indicated that the
recovery was so low that the bank could not cover
even the principal of the loan.

Information on resource requirements for the administration of the loans was not available from the
files. It was suggested however, that a good proxy
for the administration cost of maintaining a loan on
the books will be one day of loan officer time per
month during the duration of the loan. Hence for
non-insured rice loans of class one for example, the

administration cost will be given by 13.20 man-days of which 12.00 would be charged in year t and 1.20 in year t+1. A fixed cost was also suggested for the issuance of each loan; this however, varies according to the size of the loan. In addition, there could be differences in costs by items and by regions. The specific costs used in the model are discussed in the following section.

The available data did not allow estimation of a functional relation between loan supervision and loan recovery. It was indicated that each loan officer handles 60 new loans, during the expected maturity. When a loan becomes overdue it is handled by a loan recovery officer. Their job is limited to chasing the farmer around and trying to get him to pay back the loan. However, there is little the bank can do to enforce loan collection by legal means.

Financial, Physical, and Institutional Constraints

The bank faces a number of constraints of a financial, physical, and political nature regarding its institutional design. All these constraints contribute, and to great extent determine, the current structure and performance of the bank and its perspectives for growth. These constraints are taken as the initial conditions in the model discussed in the next section. Nevertheless, a number of assumptions are made regarding their changes over the planning horizon for the bank. The sources of funds are the following:

Source	1,000 US$
Internal Resources	
Loan Recovery	23,809
Interest Earnings	3,747
Government Subsidy	2,943
Other Resources	2,369
External Resources	
Borrowings	
CB's	21,400
IDB	5,915
WB	0,780
AID	1,179
Operating Budget	4,416

Besides the monetary input that these financial resources represent in the first year of the planning horizon, it is also necessary to specify the repayment conditions on borrowings and these are discussed in the following section. The operating budget for the bank is determined a priori by the board of directors, and hence it is also a condition that will influence the bank's allocation of resources. In the first year the upper limit on these expenditures was as follows:

	1,000 US$
Salaries and Honorariums	3,572
Other Operating Expenses	1,517
Capital Disbursements	214

As far as resource needs and availability, it is important to recall the seasonality in loan demand. The great majority of agriculture is rainfed and the rains begin in April. Therefore, the plantings for most crops are initiated at this time. As it can be expected, the issuance of loans is heavier during the second and third quarters. Due to bottlenecks the bank can not approve and issue all loans previous to the planting season. The seasonality in loan demand in Table 4.13 shows that most loans for rice, corn, shorghum, and coffee are issued in the second and third quarters. Industrial tomatoes are grown under irrigation and as a second crop, hence their plantings begin in November. Livestock and other loans are issued throughout the year.

As far as physical constraints, it should be recognized that a specification of available physical units is likely to underestimate the seasonal bottlenecks for loan administration and supervision. The bank currently has 162 vehicles, 169 loan officers and a total staff of 640 persons. Given the seasonality of the crop cycles, the demand for these resources varies over the year. Hence, during some months these resources may be subutilized while at other times the demand would largely exceed the supply. It would be at these times when loan requests are not evaluated properly and when loan supervision is below optimal. This version of the model does not include resource requirements in a quarterly basis, but this formulation could be included.

Considering the bank as an aggregated decision unit, one oversimplifies the extent of some institutional constraints in terms of goals of the regional agencies. Nevertheless, it is possible to specify some of the aggregate policies and goals. The most important of these is the willingness to serve the small and medium commercial farmers. Within the current administration there are no specific number of small and medium loans that must be issued, nor specific volumes of credit for particular crops, as these are determined by demand. However, it was considered appropriate (reflecting the bank's development policy and previous experience) to issue at

Table 4.13   ADBP, Number of Loans Issued by Quarter, Average of the Years, 1978-1980

Purpose	I	II	III	IV	Total
Total	1,251	1,864	2,115	1,469	6,699
Rice	56	418	208	63	745
Corn & sorghum	248	252	501	124	1,125
Industrial tomatoes	123	7	3	321	454
vegetables[a]	76	144	127	231	578
Coffee	59	178	301	50	588
Cattle	343	443	653	438	1,877
Other[b]	346	422	322	242	1,332

Source: Contraloría General de la República, Panamá, 1981.

[a]Includes potatoes, onions, and other vegetables.
[b]Includes other crops and livestock, and other purposes.

least 2,000 small loans regardless of their purpose.[3] As far as a balance in the loan portfolio, the bank tries to assign as much as two thirds of the funds to crops and one third to livestock and other items. This constraint was not imposed on the model solutions.

An important constraint is currently imposed on the issuance of insured loans. This is determined by the capacity and financial policies of ISA. Currently ISA does not provide insurance for coffee and vegetables (except industrial tomatoes and onions), and the maximum coverage is 80 percent of the value of each loan. In 1980/81 ISA provided coverage for 6,806 million US$ of crop loans and 6,307 million US$ of livestock loans. Also, the number of policies that ISA can administer is limited by its available resources to pay personnel, the number of vehicles and its operating budget. In 1980/81 ISA issued 2,722 policies (1,446 for crops and 1,276 for livestock).

The bank administers separately the loans for individual and associated producers. The bank is most willing to benefit this later group, yet in doing so runs into some problems. On one hand the administration of these rather large loans implies cost efficiency and hence a desirability for the bank. On the other hand however, the experience with associated producers is not the best in terms of loan recovery. In addition, these loans are given at preferential rates. Information was not available in terms of performance of insured loans for associated producers, hence this alternative was not considered. Therefore, given the bank's political objectives, it was assumed that there is the intent to assign at least 100 loans to associated producers, but not more than 240 loans because of the number of viable borrowers.

The above constraints are considered the most important ones. They could be interpreted as the determinants of the performance of the ADBP. The incorporation of these constraints in the bank's model makes it possible to appraise their net effect on the institution's development role. Their opportunity cost (shadow prices) would reveal the limitations for optimum financial performance.

BANK: A MULTIPERIOD LP MODEL FOR THE ADBP

This section discusses the actual coefficients used in the model for the ADBP. These coefficients are elaborated using the information in the previous section to structure the model in matrix form. The discussion refers first to the coefficients for one planning period, thus beginning by the identification of rows and columns of the model in one year. A separate section is devoted to discussing the calculation of the risk measures in the loan portfolio. The last section explains briefly the intertemporal linkages. It should be mentioned at this point that the technical coefficients (except for the right hand sides and bounds, and entries in the objective function) are equal for all years in the model. However, the submatrix for the first year includes some additional vectors to account for outstanding loans and borrowings. These borrowings impose heavy financial obligations on the bank during the second year of the planning horizon. At the same time beginning in the second year, the bank does not receive any soft loans from international financial agencies. The conjunctive effect of the above conditions, as it will be seen in the results of the model, puts the bank at a critical point in the second year of its planning horizon.

The structure of the model is only one, i.e. there is only one version of the model. However, there are two forms of the model. The first form (A) describes the current structure of the bank i.e., the vectors for demand and time deposits, and investment activities on bonds are fixed equal to zero through bounds on the corresponding columns. In the second form (B) the bounds on those vectors are released.

The Matrix for One Planning Period

The Model is structured with reference to the sources and uses of funds as shown in Table 4.10. Each entry in Table 4.10 becomes an activity or a group of activities in each annual transactions matrix. This annual transactions matrix for year one for example, includes the following vectors:

Sources of Funds

TLR00.01	,	transfer of net loan recovery funds from the previous period
TIE00.01	,	transfer of investment earnings from the previous period
BGS...01	,	government subsidy
BOR...01	,	other resources[4]
BCB...01	,	borrowings from commercial banks
BIB...01	,	borrowings from the IDB
BWB...01	,	borrowings from the World Bank[4]
BAI...01	,	borrowings from the USAID[4]
DSA01.01	,	deposits as savings accounts of 1,000 dollars

DSA02.01	, deposits as savings accounts of 10,000 dollars
DSA03.01	, deposits as savings accounts of 100,000 dollars
DCH01.01	, deposits as checking accounts of 10,000 dollars
DCH02.01	, deposits as checking accounts of 100,000 dollars

Uses of Funds

COE...01	, operating expenses
CPN...01	, office staff costs
CLT...01	, loan officer costs
CCT...01	, collection officer costs
ABF...01	, amortization of borrowed funds
AIP...01	, amortization of interest payments
LRI01.01	, loans for rice (RI), not insured (0) of less than $1,000 (1), and
.	other 27 classes of Loans
.	
.	
IDB01.01	, investment on bonds of 1,000 dollars maturing in one year
IBD02.01	, investments on bonds of 1,000 dollars maturing in three years
TNR...01	, net balance at the end of period.

It should be noted that the form A of the model, which reflects the current organization of the ADBP, fixes equal to zero the vectors for savings and check- ing accounts and investments on bonds. Therefore, the bank acts as a specialized lending agency. In both forms of the model there is no constraint on borrowings from commercial banks and after the first year there is no allowance for borrowings from inter- national development financial agencies after the first year. In both forms of the model the borrowings from the IDB, the World Bank, the AID and the Commercial Banks in year one are fixed, as they reflect the obligations already acquired. The value of all the other vectors in the model is determined endogenously.

A group of activities is defined in physical units, i.e. number of checking accounts, savings accounts, loans and bonds. Therefore, the model could choose a number of assets from alternative denominations, maturities and risk trying to make the best use of its scarse resources. Other activities are pure financial transactions and hence they are specified in total dollar amounts. These include

transfer of funds, borrowings from commercial banks,
operationg expenses and amortizations.
   The model's equations include a number of financial, physical and institutional constraints and the
risk balance equations. The first equation in the
model is the objective function to be maximized. In
period one, as in the other time blocks, the following equations are distinguished:

Financial Constraints

EDP...01	disposable funds
EOE...01	operating expenses
ABF...01	amortization of borrowed funds
AIP...01	amortization of interest payments
ELR...01	loan recovery
ENR...01	net returns
RLE...01	leverage requirement

Physical Constraints

RPN...01	Office personnel time requirements
RVH...01	Vehicle requirements
RLT...01	Loan Officers' time requirements
RCT...01	Loan collectors' time requirements

Institutional Constraints

RL1...01	Number of small loans (1)
RL2...01	Number of medium loans (2)
RL3...01	Number of large loans (3)
RIC...01	Value of insured crop loans
RIL...01	Value of insured livestock loans
RIP...01	Number of insured policies

Risk Constraints

RRI01.01	Return deviations in year 1 (1974)
.	.
.	.
.	.
RRI07.01	Return deviations in year 7 (1980)
RRIT0.00	Total mean of return deviatiations over the planning horizon.

Several other equations could have been included
for accounting purposes. However, it was planned
that a report writer would be used to present the
most important information, but this report writer
could not be obtained[5]. The column vectors and
equations are repeated for all years. For a 10-year
planning horizon the model has 512 column vectors and
298 equations. The latter include those equations to
collect disbursed funds maturing after the tenth year.
As a general rule, all entries in the matrix which
represent an inflow of capital (supply) are designated with a minus (-), and all entries which represent an outflow (demand) of capital are designated
with a plus (+). In the objective function costs are
negative (-) and returns are positive (+).

Returns and resource requirements for the various
loan and investment activities were calculated according to the following criteria:

a.   There is an issuance cost for each loan.
This cost varies according to the size of the loan in
the following way:

Loan Class	$/loan
Small (1)	100
Medium (2)	200
Large (3)	300

This issuance cost increases with the loan size because, as the loan is for a larger amount, it demands
more time to be evaluated and higher level staff participation before the disbursement is authorized.

b.   The requirements of loan officer's time are
estimated considering that a loan officer is in
charge of a loan during the expected maturity period.
When the loan becomes overdue it passes to the collection department. As the expected maturity is calculated in months, the requirements of loan officer
time are specified in man-months. For example, in
the case of small-not insured-rice loans, the requirement is 7.85 loan-man-months (see Table 4.11).

c.   The requirements of collection officers' time
are expressed as the difference between expected maturity and actual maturity. In the above case, therefore, this requirement is 5.35 man-months (13.20-7.85
= 5.35)

When a loan has an expected maturity and/or a
collection period of more than 12 months, then the
requirements are charged as follow: 12 months are
charged in the year when the loan is issued and the
remaining is charged in the following or the two
following years, if necessary.

d.   Loans, bonds, checking and saving accounts
require office personnel for administrative purposes
in the following way:

	Man-months
Loans	1 man-month per month of actual duration
Bonds	0.1 man-month per month of actual duration
Checking accounts	3.0 man-month per year
Savings accounts	1.5 man-month per year

e. With regard to the risk component, the deviations of loans are modeled in the time year when the loan is collected. For example, if the loan's actual maturity is 9 months, then the deviations are modelled in the year of issuance. But, if the maturity is 23 months the deviations are modelled in year t+1, as this is the time when returns are collected (see Figure 4.1). Bonds are assumed to be riskless assets.

f. The initial conditions and the repayment of outstanding debts is shown in Table 4.14. These vectors enter as fixed (with bounds) in all solutions of the model as they reflect obligations acquired previously to period one.

g. Leverage requirements for all assets and liabilities are calculated according to the criteria in Chapter 3. The variable for net worth was approximated by the endogenously generated end of period Net Returns (see row vector RLE...01 and the column vector TNR...01 in Figure 4.1[6]

h. The summary of constraints in year one of the planning horizon included:

Number of loans to small farmers	$\geq 2,000$
Number of loans to associated producers	$\geq 100, \leq 240$
Number of insurable loans	$\leq 2,722$
Value of insurable crop loans (1,000 US$)	$\leq 6,806$
Value of insurable livestock loans (1,000 US$)	$\leq 6,307$
Office personnel (loan man-months)	$\leq 116,160$
Loan officers time (loan man-months)	$\leq 71,520$
Collection officers time (loan man-months)	$\leq 14,400$

These constraints are assumed to be enlarged during the planning horizon at rates observed in the past. The magnitude of the constraints after year one is discussed later on in this chapter.

Table 4.14
ADBP, Initial Financial Conditions in the Model (million US$)

Item	Year	Loan Recovery	Government Subsidy	Commercial Banks	Borrowings			Other	Total
					IDB	World Bank	AID		
Sources									
Borrowings	01	–	–	21,400	5,915	0,780	1,179	2,362	31,636
Loan recovery	01	23,809	–	–	–	–	–	–	23,809
Collection of Interest	01	3,747	–	–	–	–	–	–	3,747
Others	01	–	2,943	–	–	–	–	–	2,943
Uses									
Amortization of borrowings	01			7,133				0,592	7,725
	02			7,133				0,592	7,725
	03			7,133	1,478			0,592	9,203
	04				1,478	0,195		0,592	2,274
	05				1,478	0,195	1,179		2,852
	06				1,478	0,195			1,673
	07					0,195			0,195
Interest payments on borrowings	01			2,568	0,354	0,039	0,047	0,118	3,126
	02			1,712	0,354	0,039	0,047	0,088	2,240
	03			0,856	0,266	0,039	0,047	0,059	1,267
	04				0,177	0,029	0,047	0,029	0,282
	05				0,088	0,019			0,107
	06					0,009			0,009
	07					0,001			0,001

Source: Tables 4.10 and 4.11 and unpublished information.

Risk Measures in the Model

Given the objective of this research, particular interest was given to measure risk in the loan portfolio and its impact in managing the bank. Also, given the design of the model, it was necessary to take account of lending risks in the model's objective function. Therefore, prior to discussing the aggregate structure of the ADBP model, this section explains the measures of risk in the loan portfolio.

Information on loan returns was obtained for seven years (1974-1980) for non-insured loans and for the last three years for insured loans. The calculations are given in Table A.1 in the appendix. Because the model includes mean absolute deviation of returns the data had to be adapted. This was necessary because for a given loan class (let's say loans of less than $1,000, insured for rice) the average disbursed amount was different for every year.

The procedure followed is discussed with reference to Table 4.15. This is the case of non-insured rice loans of less than $1,000. The first seven columns contain information from the survey. The last row provides the column averages. Columns (8) through (11) are calculated as follows.

col (8), the monthly actual rate of interest in year s (s = 1...7) is computed as:

$$\text{col}(8) = \frac{\text{col}(7)/\text{col}(6)}{\text{col}(2)/100}$$

col (9), the actual rate of interest for the average maturity is:

$$\text{col}(9) = \text{col}(8) \times \text{col}(\bar{6})$$

where col ($\bar{6}$) is the average actual maturity (average of column (6) during the observation period (s=7).

column (10), the size adjusted actual loan collection is:
$$\text{col}(10) = \text{col}(2) \times (1.0 + \text{col}(9)/100)$$

column (11), the deviation from the average size adjusted loan collection is:

$$\text{col}(11) = \text{col}(\bar{10}) - \text{col}(10)$$

where col($\bar{10}$) is the period average size adjusted loan collection.

Therefore, with reference to Table 15 for year 1:

Table 4.15
Risk and Return Characteristics of Rice Loans, not Insured for Small Producers

(1)	(2)	(3)	(4)	(5)	(6)	(7)	(8)	(9)	(10)	(11)
Year	Amount Disbursed	Nominal Interest Rate	Amount Collected	Expected Duration	Actual Duration	Net Interest	Monthly Actual Rate of Interest	Actual Rate of Interest for actual Duration	R	d
1974	334	8.14	397	12.43	23.57	63	0.800	10.56	496	8.5
1975	330	8.00	360	7.00	19.00	30	0.478	6.31	528	40.5
1976	445	9.00	461	7.33	17.00	16	0.211	2.78	461	-20.5
1977	457	8.60	478	9.80	9.20	21	0.499	6.59	478	- 9.5
1978	160	9.00	168	6.67	9.66	8	0.517	6.82	479	- 8.5
1979	515	10.00	545	8.50	11.010	30	0.529	6.98	480	- 7.5
1980	900	12.00	919	3.00	3.00	19	0.704	9.29	491	3.5
x	449	9.25	475	7.82	13.20	27	0.534	7.05	487.5	0.0

Source: Sample of Loans. See Appendix A.

$$\begin{aligned} \text{Monthly Actual} \\ \text{Rate of Interest (8)} \end{aligned} = \frac{63.0/23.57}{334.0/100.00} = 0.800\%$$

$$\begin{aligned} \text{Rate of} \\ \text{Interest for Actual} \\ \text{Duration (9)} \end{aligned} = 0.800 \times 13.2 = 10.56\%$$

$$\begin{aligned} \text{Size Adjusted} \\ \text{Actual Collection (10)} \end{aligned} = (449) \times (1 + 0.1056) = 496 \text{ US\$}$$

$$\begin{aligned} \text{Deviations} \\ \text{from Size Adjusted} \\ \text{Collection (11)} \end{aligned} = 496.0 - 487.5 = 8.5 \text{ US\$}$$

The elements in the last columns are the mean absolute deviations introduced in equation (47). These deviations are entered in the model according to the procedures described in chapter 3. The correction factor (Δ) to transform the mean of absolute deviations in a matrix of standard deviations for s=7 is

$$\Delta^{\frac{1}{2}} = [\frac{T \; \pi}{2(T-1)}]^{\frac{1}{2}} = [\frac{(7.0)\;(3.1416)}{2(7-1)}]^{\frac{1}{2}} = 1.35$$

which enters the model as in Figure 4.1, $\left[\dfrac{\Delta^{\frac{1}{2}}}{s}\right]^{-1} = 5.18$

At this point it should be recalled that according to the discussion in chapter 3, the intertemporal linkage in the risk of return is established through the correlation on bank returns between time periods. The correlation of loan recovery between time periods was calculated equal to 0.303. But since the bank receives yearly subsidies to make up for losses, the total available funds for lending, or in other words the net balance at the beginning of each period, grows more steadily. This is evidenced by the high correlation of total available funds over time which was calculated equal to 0.807. Given that this value is close to 1.00, it was assumed that the standard deviation of returns for the whole planning horizon was equal to the sum of standard deviation of returns of each period in the horizon. Therefore, the model did not have to include intertemporal linkages to measure covariance effects among the matrices of mean absolute deviations.

The Planning Horizon and the Intertemporal Linkages

The model is structured for a 10-year planning horizon. This length was chosen as sufficient to

Figure 4.1 Sample of Vectors of the Model Showing Intertemporal Linkages.

	Sources of Funds		Operating and Financial Costs	Uses of Funds	
	Borrowing	Deposits		Loans Issued	Risk on Loans

OBJ   N

EDP...01 E
FOE...01 E
ABF...01 E
AIP...01 E
RPN...01 L
RVH...01 L
RL1...01 G
RL2...01 G
RL3...01 G
RIC...01 L
RIL...01 L
RCP...01 G
RLV...01 G
ELR...01 E T
ELB...01 E
ERR...01 E
RLE...01 E
RLT...01 L
RCT...01 L

```
RIP...01 L
RRI01.01 G          V-U-T    1      V-V-V-U              U      1
RRI02.01 G          U-U T           V -V-U              U      11
RRI03.01 G         -U U-T          -U-U-V-U            -U     -T      1
RRI04.01 G         -V U-T          -V-U -V             U-V-U-T       1
RRI05.01 G        -V-U U...V V U V V V              U  T       1   1
RRI06.01 G        -V U T...V-V-U V U U             -U V  T       1   1
RRI07.01 G         V U T...U V U V U              V V U              1

EDP...02 E                                                          V
BOS...02 E     C D
ABP...02 E     B U C B B
AIP...02 E              A           T T B B A A A     A
RPN...02 L

EIR...02 E                                          -A-T
EI3...02 E                                          -A-T
ENR...02 E
RLE...02 E
RLT...02 L                          D B A A         A
RCI...02 L
RIP...02 L
RRI01.02 G                                          -V
RRI02.02 G                                          -U
RRI03.02 G                                          -U
RRI04.02 G                                           V
RRI05.02 G                                          U U
RRI06.02 G                                         -U V
RRI07.02 G                                           U
```

Figure 4.1 Sample of Vectors of the Model Showing Intertemporal Linkages.

Sources of Funds			Uses of Funds		
Borrowing	Deposits	Operating and Financial Costs	Loans Issued	Risk on Loans	

EDP..03 E
FOE..03 E
ABP..03 E
AIP..03 E
RPN..03 E

. . .

EIR..03 E
EIB..03 E
ENR..03 E
RLE..03 E

. . .

capture the intertemporal linkages of long term loans and borrowings. As discussed in the following section, this length proved to be sufficient for the model results to be consistent. The intertemporal linkages are established through various elements:

a. Loans issued in period t may be collected in periods t+1, t+2, or even t+3, providing returns only at this latter time.

b. Loans issued in period t would have maturities that imply that they use physical resources in periods t+1, t+2 or even t+3.

c. Outstanding debts in periods 1 ought to be paid back in forthcoming periods, as shown in Table 4.14.

d. New borrowings from commercial banks at time t are short term in nature, hence need to be paid in year t+1.

e. Bonds purchased in period t could mature at the end of period and provide returns only at maturity.

f. The most important intertemporal linkage is the collected funds in period t that are transferable as disposable funds for period t+1.

Some of these intertemporal linkages can be observed in the picture of the model for the first two planning periods, shown in Figure 4.2. A similar linkage exists for subsequent time periods.

Some assumptions were necessary about the rate of growth of specific parameters. As far as resource availability,, it was assumed consistently with previous experience, that office staff, loan officers, collection offcers, and vehicles will grow at a rate of 5 percent per annum. About the insurance program, ISA has indicated its strategy of increasing the coverage to crops and livestock at a rate of 10 percent per annum. But being aware of high administrative costs, ISA would prefer to concentrate on medium size policies, hence it was assumed that the total number of policies would grow only at a rate of 5 percent per year. It is also assumed that the bank's policy is to continue servicing the smaller farmers, hence the number of loans to small farmers would also grow at an annual rate of 5 percent.

It is recognized that these assumptions would influence the allocation of funds among the many alternatives, yet they may be preferable to an assumption of no change in these parameters. These growth rates are likely to be influenced by the government allocation of funds for development purposes. However, it is assumed that direct government subsidies to the bank will be maintained constant at 3 million US$ per year. Any changes in these assumptions are perfectly feasible as the bank authorities or the government enforce new policies.

VALIDATION OF THE MODEL

Introduction

A mathematical model is an abstraction of reality. Nevertheless, the structure and the data used should allow one to reproduce the real conditions with the highest approximation. On the other hand, in the context of the mathematical programming model used, one could impose all possible restrictions and hence reproduce lending and borrowing plans and levels of resource use that exactly resemble the current allocative criteria of the bank. Such a solution from the model could be far from the economic optimum as the levels of each vector are determined a priori. For example, the bank may decide the amount to be allocated to each crop, independently of any financial or economic criteria.

The model used here included some of the most important financial, physical and institutional constraints, but all the decisions on lending and borrowing, and hence the levels of resource use are endogenously determined. Therefore, a basic solution of the model approximates reality, but by no means could it resemble the actual dissagregated portfolio of the bank. With this caveat, it is understood that the solutions of the model provide insight about the possible effects of modifying alternative scenarios. The solutions, however, should not be taken literally as a 'crystal ball' prediction.

Before using the model for policy analysis, it was tested in terms of appropriateness of the planning horizon, and its response to the discount rate. The first test was performed to evaluate whether the planning horizon is long enough for the first years' solutions to be stable. The second test was necessary to determine the importance of the discount rate in the allocation of funds over time.

The Appropriateness of the Planning Horizon

The model was solved first for a planning horizon of 10 years and this is called the Basic Solution. To test the stability of the solutions in the first and second years, an eleventh year was added to the model. The linkages between years 10 and 11 follow exactly the same approach as for earlier periods.

Tables 4.16 and 4.17 give the main results of the model for the Basic Solution and for the one with 11 years. As expected, the model's objective function and the standard deviation of returns increase. The important result is that the solution of the model for the first two years did not change in terms of

Table 4.16
ADBP Model, Test of the Length of the Planning Horizon
(million US$)

Variable	Year	Length of Planning Horizon	
		10 years	11 years
Objective Function		226.117	257.237
Standard Deviation of Returns		3.101	3.532
	01-02	51.014	51.014
	02-03	34.078	34.078
	03-04	41.348	41.379
	04-05	58.192	58.293
Transfer of funds	05-06	39.332	59.672
	06-07	63.051	64.001
	07-08	67.146	68.137
	08-09	69.828	70.172
	09-10	95.091	72.624
	10-11	-	98.137
	01	49.792	49.792
	02	48.937	48.937
	03	30.956	31.052
	04	45.318	45.478
Total lending	05	56.201	56.834
	06	59.695	50.837
	07	63.530	65.532
	08	66.198	67.137
	09	87.259	69.930
	10	40.742	88.038
	11	-	41.737

capture the intertemporal linkages of long term loans and borrowings. As discussed in the following section, this length proved to be sufficient for the model results to be consistent. The intertemporal linkages are established through various elements:

    a. Loans issued in period t may be collected in periods t+1, t+2, or even t+3, providing returns only at this latter time.

    b. Loans issued in period t would have maturities that imply that they use physical resources in periods t+1, t+2 or even t+3.

    c. Outstanding debts in periods 1 ought to be paid back in forthcoming periods, as shown in Table 4.14.

    d. New borrowings from commercial banks at time t are short term in nature, hence need to be paid in year t+1.

    e. Bonds purchased in period t could mature at the end of period and provide returns only at maturity.

    f. The most important intertemporal linkage is the collected funds in period t that are transferable as disposable funds for period t+1.

Some of these intertemporal linkages can be observed in the picture of the model for the first two planning periods, shown in Figure 4.1. A similar linkage exists for subsequent time periods.

Some assumptions were necessary about the rate of growth of specific parameters. As far as resource availability, it was assumed consistently with previous experience, that office staff, loan officers, collection officers, and vehicles will grow at a rate of 5 percent per annum. About the insurance program, ISA has indicated its strategy of increasing the coverage to crops and livestock at a rate of 10 percent per annum. But being aware of high administrative costs, ISA would prefer to concentrate on medium size policies, hence it was assumed that the total number of policies would grow only at a rate of 5 percent per year. It is also assumed that the bank's policy is to continue servicing the smaller farmers, hence the number of loans to small farmers would also grow at an annual rate of 5 percent.

It is recognized that these assumptions would influence the allocation of funds among the many alternatives, yet they may be preferable to an assumption of no change in these parameters. These growth rates are likely to be influenced by the government allocation of funds for development purposes. However, it is assumed that direct government subsidies to the bank will be maintained constant at 3 million US$ per year. Any changes in these assumptions are perfectly feasible as the bank authorities or the government enforce new policies.

the composition of the loan portfolio, the transfer
of funds and the borrowings from commercial banks.
In fact, small changes began only in the fourth
period and they become more significant towards the
seventh period.

The above will support the proposal that a ten
year planning horizon is sufficient if the major interest is in analyzing the impact of policies over
the first two or three years of the model. However,
the above assumes that the policies evaluated do not
imply consideration of assets and liabilities of
longer maturities. With this limitation it could be
assumed that with a 10 year model, the bank can evaluate the impact of some policies four or five years
ahead of the current period.

Sensitivity to the Discount Rate

There is not perfect certainty about the correct
discount rate to be used. In the case of a public
bank, where profits are not distributed at the end of
each period, the discount rate could be ignored.
However, as this general model is applicable to banks
with different proportions of public and private ownership, the rate of discount was changed from 5 to 10
percent to test the sensitivity of the model. The
corresponding discount factors are given in Table
4.18.

If profits were distributed at the end of each
year, the expected effect of an increase in the discount rate will be a preference for earlier returns.
The above would be obtained if the bank increased the
issuance of short term loans, hence it would collect
the money earlier. Also, as insured loans in general
have shorter maturities, one would expect a preference for insuring a larger proportion of the portfolio. The latter would not necessarily imply a
larger number of insured loans, but a larger volume
of insured credit.

The data in Table 4.19 and Figure 4.3 show that
as the discount rate increases, the bank tries to
lend more in the earlier periods to get money back
earlier. In fact, at a discount of 5 percent, by
year 4 the bank would have loaned 31.89 percent of
the total, but at a rate of 10 percent, by year 4 the
bank would have loaned 33.71 percent, a slight increase over the earlier figure. Stronger adjustments
are inhibited by resource availability and leverage
requirements.

Since livestock loans have longer maturities than
crop loans, the first will decline in the first years
as the discount rate increases, as shown in Table
4.20. However, because the number of insurable loans

reaches the upper limits, the volume of non-insured credit in the first years increases. As non-insured credit is now a larger proportion of the portfolio, the amount of physical resources used by the bank increases in comparison to the situation with a lower discount rate.

The importance of the discount rate is clear, yet a careful analysis must be made of this parameter to determine the value that should be used. The above will be necessary particularly for a private bank. In the following analysis, the discount rate was maintained at the original level of 5 percent.

Table 4.17
ADBP Model, Portfolio Composition in Years 1 and 2 with 10
and 11 years Planning Horizon (Number of Loans)

	Length of Planning Horizon	
Year 1	10 years	11 years
Rice, not insured, large producers RI03	1.766.12	1.766.12
Rice, insured, large producers RI13	267.92	267.92
Rice, insured small producers RI11	2.000.00	2.000.00
Livestock, not insured medium producers LV02	202.27	202.27
Rice associated not insured RA03	100.00	100.00
Year 2		
Rice, not insured large producers RI03	959.79	959.79
Rice, insured small producers RI11	1.493.04	1.493.04
Industrial tomatoes not insured, medium producers ITU2	1.261.81	1.261.81
Vegetables, not insured, small producers VG01	606.96	606.96
Livestock, not insured medium producers LV02	2.245.89	2.245.89
Livestock, insured medium producers LV12	1.446.96	1.446.96
Rice associated, not insured RA03	240.00	240.00

Table 4.18
Discount Factor for Each Year of the Planning Horizon
at Discount Rates of 5 and 10 percent.

Year	5%	10%
1	.952	.909
2	.907	.826
3	.863	.751
4	.822	.683
5	.783	.620
6	.746	.564
7	.710	.513
8	.676	.466
9	.644	.424
10	.614	.385
11	.584	.350

Source: Gittinger, J. P. Compounding and Discount-ing Tables, John Hopkins University Press, Baltimore, 1973.

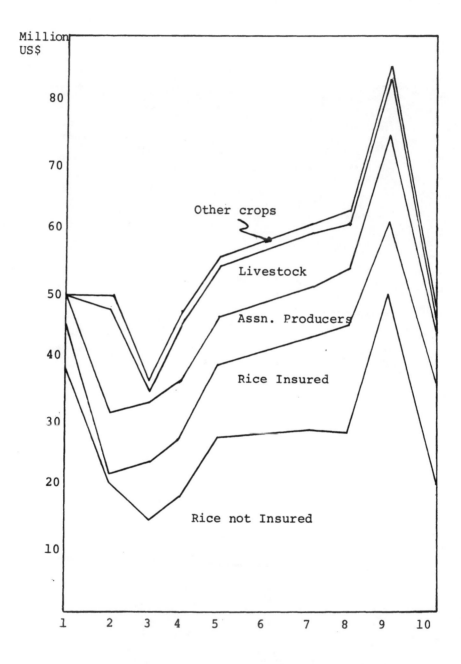

Figure 4.2 ADBP Model, Intertemporal Composition of the Loan Portfolio, 10 years.

Table 4.19
ADBP Model, Sensitivity of Lending and Transfer of
Funds to the Discount Rate (million US$)

Variable	Year	discount rate (%)	
		5	10
Objective		226.17	78.732
Standard Deviation		3.101	3.067
Transfer of funds	01-02	51.014	51.016
	02-03	34.078	37.777[a]
	03-04	41.348	42.904
	04-05	58.192	57.402
	05-06	59.332	58.567
	06-07	63.051	62.309
	07-08	67.146	66.428
	08-09	69.828	69.134
	09-10	95.091	94.371
	Total	539.080	539.904
Lending funds	01	49.792	49.794
	02	48.937	50.407
	03	30.956[a]	32.135
	04	45.318	47.183
	05	56.201	55.465
	06	59.695	58.982
	07	63.530	62.851
	08	66.198	65.355
	09	87.259	89.498
	10	40.742	20.834
	Total	548.628	532.504

[a]The drop in availability of funds at the end of
year two is due to the commitments of the bank and
the insufficient government subsidy built in the
model. This also determines the rather small loan
issuance in year three.

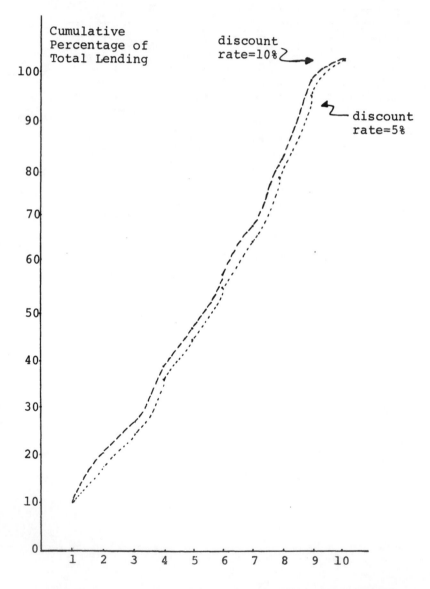

Figure 4.3   ADBP Model, Distribution of Cumulative
Lending at Different Discount Rates.

Table 4.20
ADBP Model, Maturity Structure of Loans and Prefer-
ence for Insurance at Different Discount Rates in
Years 1 to 4 (million US$)

Variable	Year	Discount rate (%)	
		5	10
Volume of credit in	1	48.986	48.986
loans of less than	2	32.697	36.558
one year	3	30.956	32.135
	4	36.493	38.369
Volume of credit in	1	0.806	0.806
loans of more than	2	16.240	13.849
one year	3	0.000	0.000
	4	8.814	8.814
Volume of insured	1	7.146	7.146
credit	2	8.163	8.163
	3	8.647	8.647
	4	18.335	18.335
Volume of not insured	1	42.646	42.646
credit	2	40.774	42.244
	3	22.309	23.488
	4	26.972	28.848

NOTES

1 ISA also insures the investment of farmers without official credit.

2 ANAGSA is the government owned and highly subsidized agricultural insurer, which serves primarily small and "ejido" farmers. SAM stands for the Mexican Food System.

3 This is to say that the small loan requirements will be met by issuing non-insured loans for all crops and livestock and insured loans for rice, corn, industrial tomatoes, and for other loans.

4 These vectors are included only in year 1. A coding for the rows and columns is presented in Appendix B.

5 The version of MPSX used (in an IBM/360-40 machine) did not include a report generator. Advanced versions of MPSX/370 that can be used in equipment of greater capacity do include report generators.

6 Leverage requirements were used at the same levels recommended by the U.S. Examinors Criteria discussed in chapter 3. Information on this issue was not available in Panama.

4
Policy Analysis
And Growth Perspectives

THE APPROACH TO POLICY ANALYSIS

The results of the model are a guide to the expected effects of various changes in policy. Therefore, the model does not produce a solution to the
problems, but only an indication of how to go about
solving such problems. This is facilitated by the
information generated in each solution for all variables in the model. This chapter discussed the main
results of a series of important changes in strategy
and policies. These changes are compared with the
basic solution of the model and with other solutions
in order to highlight the main findings.

Since the model has 298 equations and 513 activities, the information provided is extremely detailed.
This information is contained in the activity levels
and shadow prices or opportunity costs of each row
and column vectors. However, only a few variables
have been selected for discussion purposes. The emphasis on a particular variable or group of variables
changes depending upon the type of policy under
consideration.

It is convenient to recognize the scenarios for
policy analysis. In principle there is only one version of the model which is resolved under alternative
assumptions. The first three analyses involving
credit insurance and interest rate policies, the cost
of serving small farmers and the higher cost of funds
were made under the current organization and financial structure of the ADBP (form A of the model),
i.e., whithout savings and checking accounts and without possibilities for investment in the bond markets.
The following analyses were done after relaxing the
above conditions, assuming that the bank would fulfill multiple functions.

In summary a total of 11 possible scenarios were
simulated through changes in important parameters.
These were:

A.1   The basic solution; in which it was assumed that the ADBP operated under neutral attitudes towards risk ($\phi=0$) and with the physical, financial and institutional constraints at the levels discussed in chapter 4.

A.2   It was assumed extreme risk aversion ($\phi=-3.15$) with the same constraints as in A.1.

A.3   It was assumed a neutral attitude towards risk ($\phi=0$), without participation in the agricultural credit insurance program.

A.3a   It was assumed the same conditions as in A.3, but with a 2 percent increase in interest rates on all loans.

A.4   It was assumed extreme risk aversion ($\phi=-3.15$) without participation in the agricultural credit insurance program

A.5   It was assumed the same conditions as in A.1 but without a requirement for making small loans.

A.6   It was assumed the same conditions as in A.5 but with an increased cost of borrowed funds from commercial banks from 6.0 to 8.0 percent annual rate of interest.

A.7   It was assumed that besides a higher cost of funds (A.6) the bank will operate without government subsidy.

B.1   In this case the bank was assumed to have the opportunity for investment on bonds and issuance of checking and saving accounts, with all other conditions as in the case of A.7

B.2   It was assumed that the bank also had investment opportunities like in B.1 but with an increase in the annual interest rate on borrowed funds from 8.0 to 10.0 percent.

B.3   It was assumed that besides a higher cost of funds like in B.2, the bank will increase the interest rate on loans by 2 percent (as in A.3a).

The above solutions were chosen for discussion purposes and to illustrate the usefulness of the model in evaluating an alternative hypothesis. Many other solutions are possible under alternative scenarios.

DEVELOPMENT BANKING AND CREDIT INSURANCE

Bank Attitudes Towards Risk, Interest Rate Policies, and Credit Insurance

Decision makers in general show varying degrees of risk aversion in the sense that they have different preferences for return relative to the variance of return. Bankers are no exception, although the type of bank ownership will influence the attitude towards risk.   It is assumed that public

institutions are less concerned with risk management,
as any significant losses can be recovered through
government allocations. In addition, public institutions have been able to reduce financial risk by obtaining low-cost long-maturity funds from international financial agencies. Private lenders are likely to be more concerned with risks because they do
not have access to free government funds and because
they borrow almost exclusively in the commercial
capital market.

This section illustrates how credit insurance
benefits a bank serving the agricultural sector, even
when there is a neutral attitude towards risk. To
start, we should recall the hypothesis that credit
insurance has three effects, first it improves average loan recovery; second, it reduces risk of returns;
and third, it diminishes administrative costs. But,
it has also been mentioned that within a srategy of
trading risk by return, the bank could increase interest rates to compensate for low repayment. Hence,
the higher returns to the bank (because of higher
rates) would be an incentive to afford the existing
low recovery without demanding credit insurance. The
issue is, then, by how much would interest rates have
to increase to produce benefits to the bank, which
are comparable to those of credit insurance?

The information in Table 5.1 shows the main
growth indicators when the bank is risk neutral ($\phi = 0$)
and when the bank is extremely risk averse ($\phi = 3.15$).
The expected growth paths for a bank operating under
a risk neutral or an extreme risk averse type of
management were found to be very similar. However,
when the bank acted under a risk averse manner, the
path of growth was more stable and total utility over
the planning horizon, was slightly smaller.

The above suggests that the variance of returns
in the loan portfolio is rather small and hence there
would not be a significant change in the allocation
of funds as the degree of risk aversion increases.
This condition in turn implies that average recovery
is the determinant factor for choosing among various
types of loans.

The benefits of insurance are analyzed under a
situation of risk neutrality, therefore; since loans
will be issued without risk considerations, the
results will show the largest benefits of insurance.
For the purpose of this analysis, it was asumed that
the bank could issue only loans without insurance
(see Table 4.11). When insurance is not available,
loan recovery declines; however, given the bank's
leverage requirements, borrowings from commercial
banks also decline but not as much. The net effect
is a decline in lending activity as the net transfer

of funds between periods also diminishes, as it is illustrated in Figure 5.1 for the risk neutral case with and without insurance.

The opportunity cost of the restriction on the number of insurance policies declines over time because of the assumed growth rate of ISA's program (see Table 5.2). In spite of this latter assumption, all shadow prices suggest that it would pay to the bank that ISA provides with a larger number of policies. For example, one additional insurance policy in year two would increase the utility of the risk neutral bank by US$5,190 over the 10 year horizon.

The ADBP model results in Table 5.3 show a definite bank preference for insuring loans for rice and livestock. Arcia (1982) reports that these two items are the ones in which ISA has the lowest administrative costs. These items also dominate ISA portfolio as described in Arcia's model. It is surprising, however, that the bank model does not show a preference for insuring loans for corn and sorghum, the items in which ISA has had the largest losses. The reason may be that, besides being very risky, loans for corn and sorghum are unprofitable for the bank in terms of average recovery. Similarly, loans for industrial tomatoes appear more attractive because of large returns, but these loans tend ·to be rather small, and therefere; from a resource use point of view, loans for tomato production are not appealing to the bank.

There is a clear indication that insurance has a net positive effect on bank growth. This net effect is however, the result of a number of forces. The availability of insurance, even when it is only for a portion of the total loan portfolio, allows for a larger number of medium size loans, which are also of shorter actual maturity than large loans. Hence, money is turned around more rapidly. In addition, as shown in Table 5.4 loan recovery becomes a more significant proportion of internal resources; the bank's leverage position increase and therefore it is able to borrow a larger amount; the expenditures on loan supervision and collection, as a proportion of total uses of funds, is reduced, therefore leaving a larger availability of loanable funds. It should also be observed in Table 5.5 that when insurance is available the total volume of loans issued increases and the average administration costs decline. The results also show that the average size per loan declines when insurance is available, therefore; reflecting the bank's increased capacity to serve smaller farmers.

Insurance shows as a very rewarding policy for the bank. However, insurance is costly to administer

Table 5.1
BDA Model, Effects of Risk Aversion Without Insurance on
Some Growth Indicators (million US$)

Variable	Year	Risk Neutral	Risk Averse
Objective Function	10 years	196.424	181.919
Standard Deviation		4.264	3.924
	01-02	51.161	51.169
	02-03	22.353	33.930
	03-04	38.725	43.591
	04-05	53.107	50.633
Transfer of funds	05-06	53.308	50.913
	06-07	56.031	53.708
	07-08	58.986	56.758
	08-09	61.698	59.190
	09-10	66.334	63.557
	Total	461.699	463.449
	01	49.603	49.629
	02	37.941	42.541
	03	36.510	40.199
	04	32.904	38.773
Total lending	05	51.175	48.873
	06	53.789	51.559
	07	56.628	54.488
	08	59.233	56.824
	09	63.686	60.902
	10	35.702	34.221
	Total	477.171	478.009
	01	21.400	21.400
	02	0.413	6.065
	03	23.115	21.420
	04	20.498	19.468
Borrowings from	05	21.630	20.633
commercial banks	06	24.066	23.099
	07	25.460	24.544
	08	26.510	25.287
	09	29.476	28.089
	Total	192.568	190.005
Solution		A.3	A.4

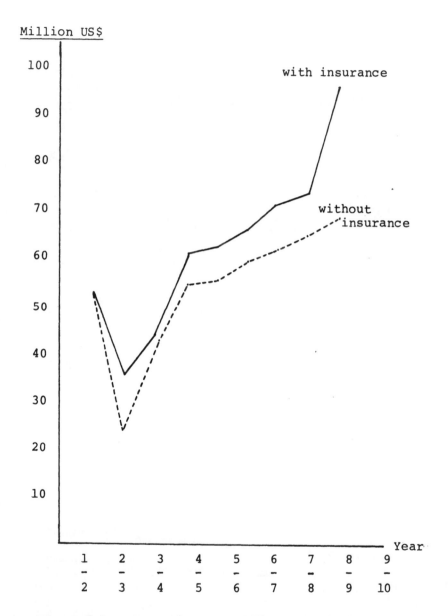

Figure 5.1   ADBP Model, Transfer of Funds Between
Periods, Risk Neutral Case.

Table 5.2
ADBP Model, Total Number of Loans Issued and Opportunity Costs of Insurance (10 years).

Variable	With Insurance Risk Neutral/Risk Averse		Without Insurance Risk Neutral/Risk Averse	
By Size				
Small[a]	25,156	25,156	25,156	25,156
Medium	18,663	22,878	7,051	8,966
Large	18,331	18,161	19,127	26,063
Total	62,150	66,195	51,334	60,185
Insured Loans	33,566	33,566	0	0
Shadow Prices of the Number of Insurance Policies (1,000 US$/policy)				
1	0,000	0,000	–	–
2	5,190	4,748	–	–
3	0,000	0,000	–	–
4	0,316	0,291	–	–
5	0,273	0,240	–	–
6	0,233	0,202	–	–
7	0,194	0,166	–	–
8	0,155	0,130	–	–
9	0,118	0,133	–	–
10	0,000	0,000	–	–
Solution	A.1	0.2	A.3	A.4

[a]The number of small loans was always equal to the minimum requirement i.e. 2.000 loans in the first year and a growth rate of 5 percent per year.

Table 5.3
Comparison of the ADBP and ISA Insurance Preferences

| Variable | Total | Insured Items | | | | | |
		Rice	Corn Sorghum	Industrial Tomatoes	Livestock	Other
Actual ISA's[a] coverage (million US$)	13,118	3,340	2,116	1,289	6,329	0,044
ISA's administrative[a] costs as a percentage of coverage	–	2.54	8.96	11.59	3.70	11.68
ISA's coverage in[a] Arcia's Model (%)	13,118 (100.0)	3.725 (28.4)	1.574 (12.0)	1.181 (9.0)	6.559 (50.0)	0.079 (0.6)
ISA's coverage in[b] the ADBP model (million US$)	14,133	6,849	–	–	7,284	–
Basic solution for[b] year 2 (%)	100.0	48.5	–	–	51.5	–

Sources: [a]Arcia, (1982).
[b]Results from the ADBP model.

and as shown in Table 5.6, the average administration costs of ISA are not diminishing. The average cost per loan for the bank shows a declining tendency and it is likely that the insured portion of the bank portfolio would have a much lower average administration cost. It would appear therefore, that if ISA could lower further its administration costs, the total government cost (in term of administrative subsidies) when issuing insured credit will be smaller than the cost when issuing not insured credit.

Insurance seems to provide more benefits to the bank through a reduction in costs, than through a reduction of risk of income. The concern is therefore, whether similar benefits can be achieved by more efficient administration of loans or by increasing interest rates on loans. This latter issue was analyzed by simulating an increase of interest rates by 2 percent on all loans (like from 9 to 11 percent per year). This percentage change is much smaller than the average insurance premiums of 5.5 percent paid by farmers.

The results of the proposed change (shown in Tables 5.4 and 5.5) indicate that on the aggregate this increase will provide benefits to the bank comparable to those of insurance of 30 percent of the portfolio. However, the risk faced by the bank will be larger as suggested by the standard deviation of returns and the bank will tend to concentrate on larger loans. Also, the average administration cost per loan will increase. Therefore, although higher interest rates are an alternative to insurance and in the long run will imply a smaller cost to the bank and to the government, it could hurt the smaller producers to which insurance is directed.

Bank Opportunity Cost of Serving Small Farmers

As it was discussed in chapter 2, ADBs and their financial policies for agriculture have been severely questioned. On the other hand, within the existing structure of the agricultural sector in developing countries, ADBs play an important role by providing low interest credit to small farmers. These financial resources would not likely be provided by profit oriented commercial banks, for whom the administration costs are too high; hence the political justification of development banks.

Although much has been written on this issue, no empirical study has measured the opportunity costs of serving small farmers. To illustrate the opportunity cost to the bank of providing these benefits to farmers, this section uses results of the ADBP model. The analysis is based on a comparison of growth paths, loan issuance and net utility in the basic model

Table 5.4
BDA Model, Sources and Uses of Funds under Insurance and Interest Rate
Policies in an Average Year under Neutral Risk Aversion[1]

Variable	Without Insurance	With Insurance	With two percent Increase in Interest Rates
		Sources of Funds	
Internal Resources			
Loan recovery and interest earnings	52.64	58.86	58.91
Government subsidy	2.99	2.99	2.99
Other resources	0.24	0.24	0.24
Sub-total	55.87	62.09	62.14
External Resources			
Borrowings from commercial banks	19.26	22.87	22.52
Borrowings from IFA's	1.79	1.79	1.79
Sub-total	21.05	24.66	24.31
Total	76.92	86.75	86.45
		Uses of Funds	
Operating Costs			
Salaries	5.73	6.10	6.23
Other costs	0.17	0.13	0.12
Sub-total	5.90	6.23	6.35
Financial Costs			
Amortization of borrowings	20.84	24.29	24.14
Interest payments	2.76	3.16	3.14
Sub-total	23.60	27.45	27.28
Loan Issuance	47.40	53.10	52.78
Total	76.90	86.78	86.41

1/This average year was calculated by dividing the total values in the model
by ten years in the planning horizon.

Table 5.5
BDA Model, Other Economic Indicators of Insurance and Interest Rate Policies under Neutral Risk Aversion (total of 10 years)

	Without Insurance	With Insurance	With two Percent Increase in Interest Rates
Objective function (mill. US$)	196,420	226,177	224,843
Standard dev. of returns (mill. US$)	4,264	3,100	4,547
Sum of end of year balances (mill. US$)	253,575	295,367	291,421
Total number of loans	51.962	65.151	53.855
Outstanding issued			
Small[a]	25,156	25,156	25,156
Medium	4,683	18,665	3,995
Large	19,123	18,330	21,704
Average size per loan (mill. US$)	10,858	7,674	10,369
Average administration cost per loan (mill. US$)	1,135	956	1,179
Ratio of loan recovery[b] to loan issuance	1,10	1.11	1.12

[a]The model had a requirement for a minimum number of small loans and this was always binding.
[b]Including interest earnings.

Table 5.6
ADBP and ISA, Total and Average Administration Costs, 1976/77 – 1982/83

	1976/77	1977/78	1978/79	1979/80	1980/81	1981/82	1982/83
ADBP							
Total cost (mill. US$)	n.a	n.a	4.236	4.405	5.304	n.a	n.a
Number of loans Issued	–	–	5.473	5.556	8.020	n.a	n.a
Number of loans outstanding	–	–	1.350	1.642	1.667	n.a	n.a
Total number of loans	–	–	6.823	7.198	9.687	n.a	n.a
Average cost per loan ($/Loan)	–	–	621	612	547	n.a	n.a
ISA							
Total cost (mill. US$)	0.123	0.188	0.218	0.318	0.463	0.682	0.714
Number of policies issued	9	351	809	2.114	2.722	3.486	4.140
Average cost per policy ($)	c/	c/	c/	151	167	196	173

a Estimated as 30 percent of loans issued in the previous year.
b Includes crop policies and livestock policies. In the second case it includes outstanding policies issued in the previous year.
c Pilot years of the insurance program.

Source: ISA, Memoria Anual, various issues
BDA, Informe Anual, Various issues

Note: When calculating the average cost it would seem unfair to use all the costs of the ADBP and those of ISA, and dividing them by the number of loans and insurance policies respectively. However, the bank does not fulfill any other function than lending and ISA does not do anything else but insurance. Hence all the expenditures of each intitution are made towards that end.

solution with the one in wich we eliminate the requirement of a minimum number of small loans.

Small loans are, in general, Less risky than large loans and they have larger average rates of return. The reason the bank rejects them stems form their relatively high cost per dollar lent. Because of the latter, profit oriented institutions such as commercial banks would prefer not to issue small loans. However, if important risk differences were to exist between small and large loans; small loans may become more desirable for a risk averse institution. As shown in Table 5.7, demanding the issuance of small loans has a negative impact on the bank's total utility, which diminished by approximately 10 percent. Operating costs, particularly the cost of loan officers and collection officers' time will be the determinant factor of lower total utility.
Table 5.8 illustrates that with the requirement of issuance of small loans, total loan officer and collection officer time increase by million US$15.64. Total utility declines by million US$21.45.

In this study the benefit can only be measured in terms of the number of beneficiaries of this policy.[1] As shown in Table 5.9, assuming a perfectly elastic demand for small loans, 30 percent more farmers would have received credit, and the average loan size would be US$8,900 as compared with an average loan size of $14,500 when small loans were not required.

The bank is able to grow at a faster rate when small loans are not required because operating costs are reduced and funds are released for loan issuance. When small loans are not obligatory, the bank would not issue any of them and the demand for loan officers' time would decline by 14 percent, and the demand for collection officers' time by over 50 percent. It should also be noted that without small loans requirements, the insurance program makes a more valuable contribution to the bank as larger loans are riskier. However, when the bank does not issue small loans, the insurance portfolio is composed of a smaller number of medium and large policies. Therefore, the restriction on the number of insurance policies in the bank model is not effective, implying that the insurance portfolio could be administered at a lower average cost, unless farmers take insurance although they do not use credit.

An indication that small loans on the aggregate are less risky is suggested by the smaller standard deviation of the bank portfolio, shown in the last row of Table 5.7. This is noticeable even when the insurance program was in operation which could suggest that without insurance, the risk of lending to large farmers is even greater than what the results in Table 5.7 suggest.

Table 5.7
ADBP Model, Growth with and without Constraint on
Small Loans (million US$)

Variable	Year	Basic	Without Restrictions on small loans
Objective Function		219,919	241,372
Transfer of funds	01-02	51,014	50,934
	02-03	34,078	34,669
	03-04	41,348	44,206
	05-06	59,332	64,464
	06-07	63,062	68,454
	07-08	67,135	72,835
	08-09	70,012	76,049
	09-10	92,042	98,331
Total lending	01	49,792	49,981
	02	48,937	51,271
	03	30,956	33,477
	04	45,318	47,988
	05	56,201	61,127
	06	59,695	64,883
	07	63,530	69,007
	08	66,198	72,000
	09	87,259	93,305
	10	40,742	43,966
Borrowings from commercial banks	01	21,400	21,400
	02	9,904	10,230
	03	16,065	18,333
	04	22,379	24,600
	05	23,867	26,207
	06	26,680	29,150
	07	28,534	31,147
	08	29,071	31,850
Standard deviation		3,213	3,558

Table 5.8
ADBP Model, Personnel Costs With and Without Restriction on Small Loans (million US$)

Variable	Year	Basic	Without Restriction on small Loans
Loan officer time	01	2,866	2,160
	02	7,509	7,509
	03	5,335	5,189
	04	6,712	6,088
	05	5,855	4,815
	06	6,325	5,205
	07	6,839	5,634
	08	7,343	6,046
	09	8,601	7,201
	10	3,573	2,523
	Total	60,958	52,370
Collection officer time	01	0,579	0,579
	02	1,512	1,512
	03	0,214	0,253
	04	0,945	0,326
	05	1,177	0,496
	06	1,305	0,524
	07	1,446	0,554
	08	1,581	0,564
	09	2,001	0,844
	10	1,250	0,299
	Total	13,010	5,951
Solution		A.1	A.5

Table 5.9
ADBP Model, Number of Loans Issued and Average Size
of Loans with and without Constraints on Small Loans

Variable	Year	Basic	Without Constraints on Small Loans
Number of loans	01	4.336	2.434
	02	8.253	7.254[a]
	03	3.403	1.373[b,d]
	04	5.519	3.387
	05	6.270	4.127
	06	6.697	4.447
	07	7.159	4.798
	08	7.591	5.113
	09	8.897	6.292
	10	4.721	1.847[c]
	Total	62.846	41.072
Size of loan (US$)	01	6,825	20,531
	02	5,930	7,067[a]
	03	9,096	24,383[b,d]
	04	8,221	14,168
	05	8,964	14,814
	06	8,913	14,592
	07	8,879	14,381
	08	8,720	14,080
	09	9,807	14,829
	10	8,621	43,966
Solution		A.1	A.5

[a]Mostly medium size loans.
[b]Only large loans.
[c]Only crop loans, because livestock loans have longer maturities.
[d]It should be recalled that the shortage of funds at the end of years one and two, and the financial commitments in year three, force the bank into the most constrained allocation of funds hence in year three it issues primarily large loans.

It is evident that requiring the issuance of small loans has some impact on slowing bank growth. Nevertheless, it should be recognized that small loans have expected returns that are in relative terms greater than for large loans, their average actual duration is shorter and they are less risky. Therefore, their imposition on the bank's portfolio does not imply such a significant cost in financial terms.

While this particular case is instructive, caution should be exercised not to attempt to generalize this data to other ADBs. Nevertheless, it suggests that serving small farmers may not impose a very severe burden in terms of the net opportunity cost of funds. Furthermore, it should be kept in mind that the proportion of small loans used for agricultural production, is usually greater than that of large loans. This is because the diversion of low interest credit outside agriculture is more common among large farmers with other investment alternatives besides agriculture (Gonzalez Vega, 1977; Adams, 1981). This would imply that although the banks bear a cost when serving small farmers, the contribution to agriculture is more meaningful.

FINANCIAL STRESS AND THE ALTERNATIVE OF DIVERSIFIED BANKING

Higher cost of funds and more constrained government subsidies may affect the feasibility of development banking with preferential rates to the agricultural sector. A higher proportion of funds are being acquired at commercial rates, thanks to government subsidies. But as the latter dry out the banks may not have other alternatives than diversification of sources and uses of funds and the use of higher interest rates on loans.

Higher interest rates on loans are largely advocated, yet it is possible that this policy may lead to higher rates of default. Because this information is not available, it was assumed that the rates of default would not change for small increases in interest rates. An analysis was also made of higher interest rates and the elimination of government subsidies and their impact on bank borrowing and growth.

The Cost and Availability of Funds and the Impact on Growth

An increase in the cost of funds reduces the bank's rate of growth. Figure 5.2 illustrates that an increase in the net cost of funds from commercial banks, from 6 to 8 percent, slows the bank's growth.

Table 5.10
ADBP Model, Shadow Prices of the Restrictions on
Small Loans and Cooperative Loans

Year	Small Loans 1,000 $/Loan	Cooperative Loans 1,000 $/Loan
01	0.163	-0.928 LL (gain)[a]
02	7.809[c]	5.279 UL (loss)[b,c]
03	0.279	0.762 UL
04	0.554	0.693 UL
05	0.472	0.585 UL
06	0.399	0.490 UL
07	0.328	0.393 UL
08	0.259	0.298 UL
09	0.192	0.206 UL
10	0.056	0.125 UL

[a]Lower limit ($\geq$ 100 loans)
[b]Upper limit ($\leq$ 240 loans)
[c]The shortage of funds in year two makes the restriction on the number of small loans very severe. Similar interpretation is possible for the upper limit on the maximum number of very large cooperative loans.

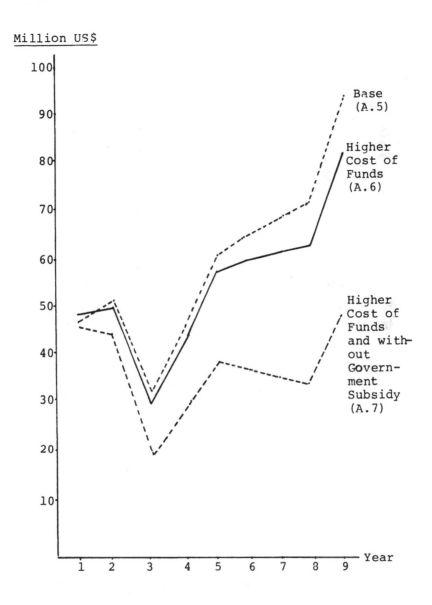

Figure 5.2   Comparative Growth Paths for Total Lending
with Higher Cost of Funds and without
Government Subsidy.

The rate of growth is reduced even further when the bank does not receive the US$3 million/year in government subsidies. In this latter case it is possible to observe a decline in total lending after the fifth year, recovering only in year nine. It is evident that a specialized ADB could not grow without government subsidies, particularly when funds from international development agencies dry out.

When the cost of funds increased, the bank adjusted its asset and liability portfolios. Table 5.12 shows that borrowings from CBs declined by 11.5 percent while the amortization of interest payments increased by 8.0 percent. On a percentage basis there was not a significant change in the source of funds. However, in the use of funds, interest payments by the bank increased from 3.64 percent to 4.31 percent, while loan issuance declined from 62.44 percent to 61.82 percent. It was observed that the size of the bank's portfolio declined, because of a reduction in loan issuance.

The elimination of government subsidies caused a more dramatic change in the structure of the bank than did an increase in the cost of funds. The size of the bank declined and total lending diminished by 61 percent. The bank became more dependent on loan recovery even though the volume of loans collected declined by 48.6 percent, while borrowings from commercial banks also declined by 54.7 percent. In terms of the uses of funds there was a small decrease in operating expenses and a further reduction in the relative importance of loan issuance which changed from 61.82 percent to 59.41 percent of the portfolio.

The loan portfolio was adjusted accordingly in response to the more expensive funds and the withdrawal of government subsidies. Table 5.13 shows that the bank would find it less profitable to issue large loans in spite of their cost effectiveness. The bank showed a preference for medium size loans which increased even when the total number of loans declined. Insurance continued to be of value to the bank as the total number of insured loans remained practically constant.

This analysis suggests that in the years ahead, a raise in the cost of funds along with a significant reduction in government subsidies would severely restrict agricultural development banking. To minimize this negative impact, ADBs would have to be more efficient in the management of their loan portfolios.

A Way Out: Diversified Banking

It is well recognized that diversification can provide one of the best strategies for bank financial management. The diversification should be allowed in the sources and uses of funds in order to generate more resources. For an agricultural development bank, lending for development purposes can be financed by generating surpluses in the bond portfolio and by attracting more funds through demand and time deposits. Within the bank's leverage requirement, these changes will contribute to enlarge the bank's capital.

To simulate a situation of diversified banking, the model was modified by allowing the issuance of bonds of one and two years maturity with nominal value of US\$1,000 and by allowing the handling of savings and checking accounts. Savings accounts could be of three average sizes: US\$1,000, US\$10,000, and US\$100,000. Checking accounts could be of US\$10,000 and US\$100,000. It was assumed that the bank could still borrow from commercial banks at a rate of 8 percent. All other constraints were held as in solution A.7.

The leverage requirement is an important constraint in determining the structure of the bank's portfolio. The data in Table 5.14 reveals that the new opportunities for the bank have a significant impact on reducing the shadow price on the leverage requirement. In fact, it could be interpreted that the new sources and uses of funds increase considerably the bank's financial performance.

The bank's external resources were primarily borrowings from commercial banks and only in two years the portfolio included checking account deposits. It is important to point out that there is no specific intertemporal linkage in terms of the number of savings and checking accounts that should be maintained. Therefore, the current formulation of the model assumes that at the end of any period the bank could close all its savings and checking accounts. Table 5.15 indicates that the model included a significant volume of deposits in years one and three. This occurred because the prior loan repayment schedule of the bank resulted in a shortage of funds and hence the need for deposits. It is assumed that the bank could attract as much in deposits as it needed.

With respect to internal resources, the bank was able to expand its loan recovery and obtain significant revenues from its bond portfolio. There was no specific restriction on the proportion of loans and bonds in the asset portfolio. On the aggregate this proportion was 1:4, however; it is important to notice a significant change in the composition of the

Table 5.11
ADBP Model, Growth with Higher Cost of Funds and Without Government Subsidy

Variable	Year	No Restriction on Small Loans	Higher Cost of Funds	Higher Cost of Funds and no Government Subsidy
Obj. Func.		241.372	219.800	145.737
	00-01			
	01-02	50.934	50.934	47.684
	02-03	34.669	34.059	26.267
	03-04	44.206	44.570	30.847
	04-05	63.079	60.076	44.473
Transfer of	05-06	64.464	60,024	40.312
funds	06-07	68.454	62.477	38.886
	07-08	72.835	65.273	38.009
	08-09	76.049	67.043	36.380
	09-10	98.331	66.238	51.266
	10-15	45.878	39.393	19.016
	01	49.981	49.882	46.953
	02	51.271	50.737	43.728
	03	33.477	31.906	20.428
	04	47.988	45.043	29.523
Total	05	61.127	56.862	37.922
lending	06	64.883	59.140	36.472
	07	69.007	61.741	35.544
	08	72.000	63.347	33.890
	09	93.305	81.687	48.084
	10	43.966	37.735	18.185
	01	31.563	31.563	31.563
	02	10.230	9.662	5.450
	03	18.337	17.113	11.482
Borrowing	04	24.600	22.591	15.167
from	05	26.207	23.570	14.491
commercial	06	29.150	25.796	15.156
banks	07	31.147	27.085	14.960
	08	31.850	27.232	13.780
	09	50.907	44.134	28.358
Standard Dev.		3.558	3.143	1.710
Solution		A.5	A.6	A.7

Table 5.12

ADBP Model, Sources and Uses of Funds with Higher Cost of Funds and without Government Subsidy for an Average Year.

Variable	No Restriction on Small Loans	Higher Cost of Funds	Higher Cost of Funds and no Government Subsidy
Sources of Funds			
Internal Resources			
Loan recovery and interest earnings	64.64 (69.47)	59.56 (69.71)	40.06 (72.56)
Government subsidy	2.99 (3.22)	2.99 (3.50)	0.00 (0.00)
Other resources	0.23 (0.25)	0.23 (0.28)	0.23 (0.43)
Sub-total	67.87 (72.94)	62.79 (73.49)	40.30 (72.99)
External Resources			
Borrowings from CBs	24.38 (26.29)	21.85 (25.58)	14.12 (25.58)
Borrowings from IFAs	0.78 (0.84)	0.78 (0.92)	0.78 (1.43)
Sub-total	25.16 (27.06)	22.64 (26.51)	14.91 (27.01)
Total	93.04 (100.00)	85.44 (100.00)	55.21 (100.00)
Uses of Funds			
Operating Costs			
Salaries	5.82 (6.27)	5.57 (6.53)	5.23 (9.48)
Other operating expenses	0.12 (0.13)	0.11 (0.13)	0.08 (0.16)
Sub-total	5.94 (6.40)	5.68 (6.66)	5.32 (9.64)

Table 5.12 (continued)

Variable	No Restriction on Small Loans	Higher Cost of Funds	Higher Cost of Funds and no Government Subsidy
		Uses of Funds	
Financial Costs			
Amortization of			
Borrowings	25.59 (27.51)	23.23 (27.20)	14.05 (26.54)
Interest Payments	3.39 (3.64)	3.68 (4.31)	2.43 (4.41)
Sub-total	28.98 (31.16)	26.92 (31.52)	17.09 (31.05)
Loan Issuance	58.09 (62.4)	52.80 (61.82)	32.80 (59.41)
Total	93.02(100.00)	85.41(100.00)	55.21(100.00)
Solution	A.5	A.6	A.7

Note: Number in parenthesis are column percentages of sources and uses of funds.

Table 5.13
ADBP Model, Total Number of Loans by Size and Insurance Class in Various Situations

Variable	Risk Neutral With Insurance	Risk Neutral Without Insurance	Without Restriction on Small Loans	With Higher Cost of Funds	With Higher Cost of Funds and Without Government Subsidy
			Total number of loans		
Small	25.156	25.156	-	-	-
Medium	18.663	4.685	19.790	19.816	20.154
Large	19.030	19.126	21.281	19.022	10.447
Total	62.849	48.967	41.071	38.838	30.601
			Insured and not insured loans		
With	33.567	0	20.067	20.067	19.965
Without	29.282	48.967	21.004	18.771	10.636
Solution	A.1	A.3	A.5	A.6	A.7

Table 5.14
ADBP Model, Shadow Price on Leverage Requirements

Year	Specialized Banking	Diversified Banking Basic Solution
1	0.841	0.155
2	3.041[a]	1.779[a]
3	0.635	0.070
4	0.433	0.086
5	0.382	0.153
6	0.324	0.206
7	0.266	0.233
8	0.209	0.032
9	0.145	0.067
10	0.094	0.026
Solution	A.7	B.1

[a]Leverage requirements on year 2 are significantly larger than in other years because of financial commitments

Table 5.15
ADBP Model, Borrowing and Deposits with Specialized
and Diversified Banking (million US$)

Variable	Year	Specialized Banking	Diversified Banking Basic Solution
	1	31,563	31,563
	2	5,450	121,550
	3	11,482	246,385
	4	15,167	254,999
Borrowing from	5	14,491	265,548
Commercial	6	15,156	276,740
Banks	7	14,960	284,526
	8	13,780	296,727
	9	28,358	328,413
	10	-	-
	Total	150,407	2,106,451
	1	-	9,147,800[a]
	2	-	-
	3	-	2,396,412[a]
	4	-	-
Demand and	5	-	-
time	6	-	-
	7	-	-
	8	-	-
	9	-	-
	10	-	-
	Total	-	248,789,000
Solution		A.7	B.1

[a]Deposits in checking accounts only. It should be
recalled that because leverage and reserve requirements, only a portion of these funds can be used in
investment and lending activities.

Table 5.16
ADBP Model, Net Returns, Loans and Bonds with Specialized and Diversified Banking (million US$)

Variable	Year	Specialized Banking	Diversified Banking
Objective		145,737	1,872,690
st. dev.		1,711	8,264
	1	36,438	41,431
	2	9,823	136,009
	3	13,525	378,547
	4	18,119	282,739
Net returns at	5	18,283	295,042
end of period	6	18,803	308,232
	7	18,514	317,673
	8	17,169	323,144
	9	33,167	363,442
	10	1,818	19,181
	Total	185,659	2,465,440
	1	46,953	50,564
	2	43,728	128,592
	3	20,428	3,625
	4	29,523	115,016
Loan issuance	5	37,922	118,439
	6	36,472	136,321
	7	35,544	140,128
	8	33,890	20,980
	9	48,084	139,210
	10	18,185	173,876
	Total	350,729	1,026,751
	1	-	86,071
	2	-	115,917
	3	-	363,677
	4	-	435,851
Security	5	-	592,973
Purchases	6	-	579,789
	7	-	538,838
	8	-	624,311
	9	-	496,610
	10	-	176,585
	Total	-	4,010,622
Solution		A.1	B.1

Table 5.17
ADBP Model, Interest Rate, Sensitivity of Borrowings, Lending and Purchase of Securities (million US$).

Variable	Rate on borrowings = 8% Basic Rate on Loans	Rate on borrowings = 10% Basic Rate on Loans	Rate on Borrowings = 10% Rate on Loans Increased by 2%
Total utility	1,872,690	1,931,827	1,943,118
Borrowings	2,106,453	2,166,818	2,184,582
Deposits	248,789,000	258,987,800	253,055,800
Loan issuance[a]	1,036,752	782.124	862,044
Purchase of bonds	3,580,759	4,010,622	3,984,284
Amortization of Borrowings	339,012	434,081	437,632

[a]See end note 2 in this chapter.

asset portfolio over the plannning horizon. This flexibility in the model should in fact reflect the institutional ability to adjust the bank portfolio over the business cycle, depending on the relative profitability of the various financial instruments.

As shown in Table 5.16, allowing multiple functions is an important way of increasing the capacity of the bank to allocate funds to agricultural loans. In this case no specific demand was imposed on the number of small loans, hence the bank's lending activity was primarily for large loans and some medium loans. Presumably, in this case the requirements of serving small farmers could be imposed without a significant detriment in the bank's aggregate performance.

The above discussion has illustrated that allowing the ADBs to fulfill multiple functions, is potentially an important strategy by which the banks can expand their portfolios and generate surpluses to finance agricultural development. However, as the cost of funds to the banks become higher, they will find it increasingly difficult to finance agricultural loans at low rates. This situation is illustrated with the data in Table 5.17, were the cost of funds from commercial banks was increased from 8 percent/year to 10 percent/year.

Two important effects of adjustment on the bank's portfolio take place as interest rates on borrowed funds increase. On the liability side the bank would increase its demand for deposits relative to borrowing. On the asset side, the bank would increase the investment in bonds and decrease the issuance of loans. Table 5.17 shows that there would be a slight increase in borrowings even when interest rates are higher. It should be recalled, however, that what the bank is looking for is the net availability of borrowed funds. Therefore, the bank must borrow larger volumes because amortization of interest payments is now a larger proportion of the uses of funds.

The alternative of increasing the interest rates on loans was also evaluated by assuming an average increase of 2 percent on returns of all loans. The results in the last column of Table 5.17 suggests that this policy will result in an increased preference by the bank to issue loans and a decline in the relative preference for bonds. The average elasticity of bank demand for loans is 5.0.[3] But as borrowings also increase (while deposits practically do not change), the substitution of bonds for loans is not as significant. In fact, the elasticity of substitution of loans for bonds was equal to 0.324, but it should be recalled that the volume of bonds is four times that of loans.

CONCLUDING COMMENTS

The analyses performed here are a sample from an infinite number of possible scenarios. In this sense, the uses of the model go far beyond what has been shown here. The real output of this work is the model itself and the analysis of particular issues of interest. But as other questions are presented by the bank's authorities, those questions can be analyzed with the aid of the model.

It should be recognized on the other hand that the current version of the model could be enriched considerably by further disaggregation, in the loan classes, in the sources of funds, in the investment portfolio, and in the number of financial, physical and institutional constraints. If necessary the model could also be disaggregated by regions or agencies, but this would considerably enlarge the size of the model.

The most important elements of the model are the loan vectors. If the model was to be improved, it should be in the specification of loan classes with alternative levels of loan supervision as this is an important determinant of loan recovery. In fact, the benefits of insurance are so significant in the model because it improves loan recovery and reduces administration costs. Yet, at least part of these benefits could be achieved through improved loan management.

NOTES

1. It is assumed that the bank issues only one loan per farmer.

2. Earlier work of the author (Pomareda, 1982.a) using an econometric model, showed different elasticities of bank demand for loans, municipal securities and treasury securities in the U.S when interest rates were rising than when they were falling.

3. This elasticity of bank demand for loans actually reflects the bank's marginal cost function of loan issuance (see Pomareda, 1982.a). It was calculated as the percentage change in loan issuance divided by the percentage change in interest rates

$$\frac{(862.44-782.128)/782.128}{.02} = 5.11$$

which means that an increase of one percent in interest rates on loans (e.g., from 9 to 10 percent per year) will increase the bank's demand for loans by 5.11 percent.

5
Summary and Conclusions

The objective of this study was to analyze the impact of various policies on the growth of ADBs. The policies with which the study was concerned referred to the effect of risk aversion in the management of funds, the requirement of credit insurance on agricultural and livestock production loans, higher interest rates on loans, measurement of the bank's opportunity cost of serving small farmers, portfolio adjustments resulting from higher cost of funds and withdrawal of government subsidies and finally the effects of transforming a specialized lending bank into one with multiple functions.

The institutional design of ADBs and the interest rate policies for agriculture were discussed. Many ADBs are specialized lending institutions and this limits their financial intermediation capacity. There are diverse views about the justification of interest rate subsidies, but it is well recognized that they introduce distortions in the rural capital markets. Because of the clientele ADBs serve, they face high operating costs. Therefore, low interest earnings and high costs determine low net returns to the bank and poor quality of service.

A generalized characteristic of ADBs is their poor loan collection performance. This is due in part to agricultural risks. Agricultural insurance and credit insurance are proposed as a means for stabilizing farm income, when its variability originates on yield instability. The effectiveness of insurance in stabilizing farm income depends on the origin of risks. Furthermore, even when income could be fully guaranteed, there could still be low loan repayment because of moral risk.

In order to evaluate the policies considered, a multiperiod linear programming model was constructed. The method used was based on portfolio theory and on

151

earlier bank portfolio models. The most interesting features of the model are the integration of financial, institutional, and physical constraints; the intertemporal linkages in assets and liabilities in the portfolio; a very disaggregated lending possibilities through 26 classes of loans; and a linearized measured of risk that captures variance-covariance effects on the loan portfolio in each period of the planning horizon. The model was applied to the case of the ADBP of Panama for a 10-year planning horizon. The loan classes were defined on the basis of a survey of 900 loans issued between 1974 and 1980. The average size, expected and actual recovery, and expected and actual maturity were estimated for each loan class.

The model was structured by 298 equations and 513 variables. The model was tested for the appropriateness of the length of the planning horizon and the sensitivity to the discount rate.

A ten year horizon was sufficient to analyze policies that do not modify the maturity structure of assets and liabilities. The model solutions were sensitive to the discount rate, which means that this model can be applied to financial institutions with varying degrees of ownership and hence different strategies for distributions of earnings at the end of each planning period.

The main conclusions of the study were the following:

a. Increased risk aversion provided a more stable growth of the bank, because although total utility decreased, it can be expected with more certainty. Increased risk aversion would become more important as financial support from the government and soft international loans diminish.

b. Agricultural credit insurance provides direct benefits for an ADB through higher average recovery, decreased variations of recovery over time, reduced administration and collection costs. These benefits increase as production risk becomes a more important factor in loan recovery. Credit insurance allows for reduction of costs and increase of earnings, both of which allow a faster growth of credit. Credit insurance provides the largest benefits when issuing large loans in the case of the ADBP, but this can not be generalized to other banks.

c. Higher interest rates on loans should be considered as an alternative to the bank to increase its returns. But this alternative does not lower administration costs neither it favors small producers, who are the clientele ADBs wish to reach.

d. Serving small farmers increased the bank's operating costs, but the net effect on slowing bank growth was mitigated by the fact that small loans are less risky and that they have shorter actual duration than medium and large loans. Their inclusion on bank portfolios should be considered on the basis of cost, return, and risk characteristics. All these factors could vary from crop to crop for the same size of loan.

e. Increased cost of borrowings from commercial domestic and international banks imposes a severe constraint for the growth of specialized ADBs, if they do not increase interest rates on loans issued. These increased costs could be afforded with government subsidies, but as the latter diminish, those banks that do not adjust their spreads will tend to disappear.

f. Moving a bank from specialized lending to multiple functions will require significant changes in institutional design and management, but it may be the only way for ADBs to exist when international soft loans and subsidies are not available. However, even in this case ADBs will need to increase interest rates on issued loans. But they could also in this case maintain separate portfolios for development oriented (low interest rate) loans and commercial loans.

These conclusions pertain to the main findings of this study. They can not be generalized to all ADBs yet they delineate issues for further research in ADBs in the developing countries and even in some developed countries where institutional design and interest rate policies fall within the general framework presented here.

Recommendations

This study has derived several conclusions, not only from the empirical analysis of Panama's ADB, but also from the general review of bibliography and analysis of financial policies, design, and performance of ADBs. On this basis the study offers a set of recommendations of direct relevance for the ADBP, but also of interest for ADBs in general.

Because the specialized nature of ADBs inhibits them from playing a more meaningful role in the supply of agricultural credit, ADBs are well advised to diversify their functions. This would allow ADBs to afford the withdrawal of government subsidies and the expected relative smaller contribution of international development financial agencies. However, diversification would imply charging higher rates on loans in order to maintain a positive spread as the attraction of savings deposits will be possible only at competitive rates. The banks should also explore the possibility of maintaining loan portfolios with different rates, to the extent that the banks (or the government) wish to continue favoring particular groups.

As an overall strategy, increasing interest rates to agriculture should be considered. This policy could allow the bank to pay higher prices for borrowed funds and hence to attract more funds for investment in agriculture and it is not likely to create a lasting decrease in the demand for agricultural credit by farmers. The shift towards rates more consistent with inflation and structure of the countries' capital markets should allow an increase in the immediate and long term availability of credit and it would allow the banks to offer a service of better quality.

ADBs that so far have been primarily government owned should be given more financial freedom. The above will allow them to take a more professional attitude towards management and hence, enforce loan selection supervision and prosecution that is not distorted by political maneuvers. Coupled with the above must be the awareness of dependence on self performance rather than bargaining ability to obtain government funds.

It has been shown that credit insurance can provide important benefits for an ADB. However, to the extent that loan defaults are due only partially to production risks, caution must be exercised in the creation of credit insurance programs, as these are not a substitute for better bank management. Furthermore, credit insurance should be favored only when its costs do not exceed the direct immediate and spillover long term effects of increased loan recovery. Insurers should also be given financial freedom to avoid indemnities made for political reasons which could be too costly to the insurer.

Credit to small farmers imposes an important cost to a bank in terms of the number of clients to be served. But the decisions about inclusion of small loans in bank portfolios should be based not only in terms of operating costs, but on the risk of recovery and maturity of the loans. It is necessary to consider these factors because if in fact small loans are not cost efficient, they may be less risky and they could have shorter maturities. Hence, on the aggregate some small loans may be preferred to large loans; particularly if the capital market limits the banks to liabilities with shorter maturities.

As far as management, ADBs have much to gain from the use of analytical tools, operations research methods and statistical analysis of their own experience. In this regard, this study has shown that the analysis of data and the use of a mathematical model with that data, is an important element to highlight the existence of problems. Furthermore, the models offer the chance for systematic thinking and certainly offer a guide for policy analysis. The use of the analytical tools among development banks makes sense to the extent that the bank authorities want to rely on professional advice to evaluate the financial cost of political decisions.

Bibliography

Adams, D. W. "Are the Arguments for Cheap Agricultural Credit Sound?" Colloquium on Rural Finance,
World Bank, EDI, Washington D.C., September 1-3,
1981.

Adar, Z., T. Agmon and Y. E. Orgler. "Output Mix and
Jointness in Production in the Banking Firm."
Journal of Money, Credit and Banking. 7 (1975):
235-234.

Aguirre, J. A. and C. Pomareda. Financiamiento del
Desarrollo Agropecuario de América Latina: Pers-
pectivas y Estrategia. ALIDE, Secretaría General,
Lima, 1981.

ALIDE. "La Tasa de Interés y la Banca de Desarrollo
ALIDE, Secretaría General, Lima, 1981.

Anderson, J., J. Dillon, and J. Hardaker. Agricultural Decision Analysis. Iowa State University
Press, Ames, Iowa, 1977.

Arcia, G. Portfolio Management and the Design of
Agricultural Insurance: The Case of Panama.
Paper presented to the IICA-IFPRI Conference on
Agricultural Risks Insurance and Credit in Latin
America. San José, Costa Rica; February 1982.

Baker, C. B., and D. J. Dunn. "Lending Rules of
Federal Land Banks and the Farm Credit Act of
1971." Agricultural Finance Review, 32 (November
1979):1-16.

Baltensperger, E. "Costs of Banking Activities-Inter-
actions Between Risks and Operating Costs." Journal of Money, Credit and Banking. 4 (1972,a):595-
611.

_____. "Economics of Scale, Firm Size and
Concentration in Banking." Journal of Money
Credit and Banking, 4 (August 1972,b):467-488.

_____. "Alternative Approaches to the
Theory of the Banking Firm." Journal of Monetary
Economics, 6 (1980):1-37. North Holland.

BANCO DE DESARROLLO AGROPECUARIO. Memoria Anual,
various issues. Panama city, Panama.

BANCO NACIONAL DE PANAMA. Memoria Anual, various issues. Panama city, Panama.

Barton, C. G. Credit and Commercial Control: Strategies and Methods of Chinese Businessmen in South Vietnam. PhD Dissertation, Cornell University, 1977.

Basu, S. K. A Review of Current Banking Theory and Practice. New York: MacMillan, 2d. ed., 1974.

Baumol, W. J. "An Expected Gain-Confidence Limit Criterion for Portfolio Selection." Management Science, 10 (1963):171-182.

Beazer, W. F. Optimization of Bank Portfolios. Lexington Books; Lexington, Massachussets, 1975.

Bernal, A.; J. Herrera, and L. Joly. "The Impact of Insurance on the Supply of Agricultural Credit: A Case of Chiriqui Province, Panama." Paper presented at the IICA-IFPRI Conference on Agricultural Risks, Insurance and Credit in Latin America. San José, Costa Rica; February 1982.

Berry, S. S. "Risk and the Poor Farmer." Paper prepared for AID's Technical Assistance Bureau, Economic and Sector Planning, USAID, Washington D.C., August 1977.

Binswanger, H. P. "Risk Attitudes of Rural Households in Semiarid Tropical India". Yale University, Economic Growth Center; discussion paper No. 275, 1978.

Bouman, F. J. A. "The ROSCA: Financial Performance of an informal Savings and Credit Institution in Developing Economies." Savings and Development, 3 (1979):253-276.

Boussard, J. M. "Time Horizon, Objective Function and Uncertainty in a Multiperiod Model of Farm Firm Growth." American Journal of Agriculural Economics, 53 (August 1981):467-477.

Bradley, S. P., and D. B. Crane. Management of Bank Portfolios. New York, John Wiley and Sons, 1975.

Buser, S. A.; A. H. Chen, and E. Ken. "Federal Deposit Insurance, Regulatory Policy and Optimal Bank Capital." Journal of Finance, 35 (March 1981).

Ccama, F.; and C Pastor. Efectos del Seguro Agrocrediticio, el Crédito y la Asistencia Técnica sobre la Producción de Papa en Bolivia. Proyecto IICA/ASBA; La Paz, Bolivia, June 1982.

Chambers, D. and A. Charnes. "Intertemporal Analysis and Optimization of Bank Portfolios." Management Science, 7 (April 1961):393-410.

Chen, J. T. and C. B. Baker. "Marginal Risk Constraint Linear Programming for Activity Analysis." American Journal of Agricultural Economics, 56 (August 1974):622-627.

Cohen, K. J. and S. Thore. "Programming Bank Portfolios Under Uncertainty." Journal of Bank Research, 1 (Spring 1970):42-61.

Cohen, K. J. and F. S. Hammer. Analytical Methods in
    Banking. Homewood, Illinois; Richard D. Irwin,
    1966.

_____. "Linear Programming and
    Optimal Bank Asset Management Decisions." The
    Journal of Finance, 22 (February 1967):147-165.

Crane, D. B. "A Stochastic Programming Model for
    Commercial Bank Bond Portfolio Management."
    Journal of Financial and Quantitative Analysis, 6
    (June 1971):955-976.

Crawford, P. R. "Algunas observaciones sobre el
    Seguro de Cosechas en Países en Desarrollo".
    Revista ICA, 14 (September 1979):209-214.

de Janvry, A. "Optimal Levels of Fertilization Under
    Risk: The Potential for Corn and Wheat Fertili-
    zation Under Alternative Price Policies in
    Argentina." American Journal of Agricultural
    Economics, 54 (February 1972):1-10.

Edgeworth, F. Y. "The Mathematical Theory of Banking."
    Journal of the Royal Statistical Society, 51
    (March 1888):113-127.

Frey, T. L. "Optimal Asset and Liability Decisions
    for a Rural Bank: An Application of Multiperiod
    Linear Programming." PhD Thesis, University of
    Illinois; Urbana, Illinois, 1970.

Fried, J. "Bank Portfolio Selection." Journal of
    Financial and Quantitative Analysis, 5 (June
    1970):203-227.

Galbis, V. " Manejo de las Tasas de Interês."
    MONETARIA, (CEMLA), 4 (July-September, 1981);
    263-302.

Gardner, B. "The Farmer Risk and Financial Environ-
    ment Under the Food and Agriculture Act of 1977.
    Agricultural Finance Review, 39 (November
    1979):123-141.

Gittinger, J. P. Compounding and Discounting Tables
    for Project Evaluation. Baltimore, John Hopkins
    University Press, 1973.

Gonzalez-Vega, C. "Interest Rate Restrictions and
    Income Distribution." American Journal of Agri-
    cultural Economics, 59 (November 1977):973-976.

_____. "Redistribution in Reverse: Cheap
    Credit can not Redistribute Income in Favor of
    the Small Farmer." Colloquium on Rural Finance,
    World BAnk/EDI (September 1981):1-3.

Green, D. D. "A Survey of Bibliography about Rural
    Household Behavior Models." ECID/SIECA, unpub-
    lished paper; Guatemala, July 1978.

Gudger, W. M. El Seguro Agrocrediticio y su Papel en
    la Promoción del Desarrollo Rural. Seminar on
    Perspectives for Agricultural Insurance in Peru,
    Lima; November 10-11, 1980.

Gunter, L. F. and F. E. Bender. " Ending Conditions for Finite Linear Programming Models based on Infinite Horizon Replacement Problems." Unpublished paper, University of Maryland, Dept. of Agricultural Economics. 1980.

Hanson, G. and J. L. Thompson. "A Simulation Study of Maximum Feasible Farm Debt Burdens by Farm Type." American Journal of Agricultural Economics, 62 (November 1980):727-733.

Hart, O. D. and D. M. Jaffee. "On the Application of Portfolio Theory to Depositary Financial Intermediaries." Review of Economic Studies, 41 (1974):129-147.

Hazell, P. B. R. "A Linear Alternative to Quadratic and Semivariance Programming for Farm Planning under Uncertainty." American Journal of Agricultural Economics, 53 (February 1971):53-62.
_____ and P. L. Scandizzo. "Competitive Demand Structures under Risk in Linear Programming Models." American Journal of Agricultural Economics, 56 (May 1974):235-244.
_____. Market Intervention Policies when Production is Risky. American Journal of Agricultural Economics, 56 (November 1975):645-649.

Hazell, P. B. R. and C. Pomareda. "Evaluating Price Stabilization Schemes with Mathematical Programming." American Journal of Agricultural Economics, 63 (August 1981):550-556.

Hester, D. and J. L. Pierce. Bank Management and Portfolio Behavior, New Haven, C.T., 1975.

Hogan, A. J. The Role of Crop Credit Insurance in the Agricultural Credit System in Developing Economies. PhD Dissertation, University of Wisconsin; Madison, 1981.

IDB. Progreso Econômico y Social en América Latina. Informe 1980-81, Washington D.C., 1981.

IICA. Rural Credit Insurance Project - Annual Report 1981 - Summary. IICA, San José, Costa Rica; December 1981. ISA. Memoria Anual. Instituto de Seguro Agropecuario. Panama, various issues, 1978-1981.

Jessup, P.F. Modern Bank Management. St. Paul, West Publishing, 1980.

Just, R. E.; E. Lutz; A. Schmitz and S. Turnovsky. "The Distribution of Welfare Gains from International Price Stabilization Under Distortions." American Journal of Agricultural Economics, 59 (1977):652-661.
_____ and R. D. Pope. "On the Relationship of Input Decisions and Risk". In J. Roumasset, J. Boussard and I.J. Singh (eds). Risk, Uncertainty and Agricultural Development. Agricultural Development Council, 1979.

Kane, J. Development Banking: An Economic Appraisal. Lexington Books, Massachussets, 1975.

Kane, E. and B. J. Malkiel. "Bank Portfolio Allocation, Deposit Variability and the Availability Doctrine." Quarterly Journal of Economics, 79 (February 1965):113-134.

Kim, C. S. and J. F. Yanagida. "A Comparison of Quadratic Programming, MOTAD, MRC-LP and DC-LP," University of Nevada, Reno, Div. of Agric. and Resource Economics, MS-136, May 1981.

Klein, M. A. "A Theory of the Banking Firm." Journal of Money, Credit and Banking, 3 (1971): 205-218.

Koropecky, O. "Risk Sharing, Attitudes and Institutions in the Rural Sector: A Critique of a Case Against Crop Insurance in Developing Countries." Robert Nathan and Associates, September 1980.

LA ESTRELLA DE PANAMA. Operante Sistema de Descuento sobre Tasas de Interés. Panama City, Panama. Tuesday, December 8, 1981.

Ladman, J. R. "The Costs of Credit Delivery, the Institutional Structure of Rural Financial Markets and Policies to serve more Small Farmers with Credit Programs." Colloquium on Rural Finance, World Bank/EDI, Washington D.C. (September 1981):1-3.

_____ and R. L. Tinnermeir. "The Political Economy of Agricultural Credit: The Case of Bolivia." American Journal of Agricultural Economics, 63 (February 1981):66-72.

Markowitz, H. Portfolio Selection. Wiley, New York. Cowles Foundation Monograph No. 16. 1959.

_____. "Portfolio Selection: Efficient Diversification of Investments." Cowles Foundation Monograph, 16 (1959), revised edition, published by Yale University Press, 1976.

Massel, B. F. "Price Stabilization and Welfare." Quarterly Journal of Economics, 83 (1969):284-289.

McCarl, B. and T. F. Tice. "Linearizing Quadratic Programs Through Matrix Diagonalization. American Journal of Agricultural Economics, 62 (August 1980):571-574.

Miller, L. W. "Using Linear Programming to Derive Planning Horizons for a Production Smoothing Problem." Management Science, 25 (1979):1.232-1.244.

Modigliani, A. and K. H. Cohen. "The Role of Anticipations and Plans in Economic Behavior and Their Use in Economic Analysis and Forecasting." Studies in Business Expectations and Planning, 4. University of Illinois; Champaign, Illinois, 1961.

Monti, M. "Deposit, Credit and Interest Rates Determination Under Alternative Bank Objective Functions." In. G.P. Szego and K. Shell (eds). Mathematical Methods in Investment and Finance. Amsterdam, North Holland Publishing Co. 1972.

Moscardi, E. and A. de Janvry. "Attitudes Towards Risk Among Peasants: An Econometric Approach." American Journal of Agricultural Economics, 59 (1977):710-716.

Olson, N. "A Multiperiod LP Model for Studies of the Growth Problem of Agricultural Firms." Swedish Journal of Agricultural Research, 1 (1971):155-175.

PANAMA. Comisiôn Bancaria Nacional. Boletines Estadîsticos, various issues, 1981 and 1982.

_____. Contralorîa General de la Repûblica. Panamâ en Cifras 1974-1978, Panama 1979.

_____. Direcciôn de Estadîstica y Censo, Contralorîa General. Panama, 1981.

Parkin, M. "Discount House Portfolio and Debt Selection." Review of Economic Studies, 37 (1970):469-497.

Pesek, B. "Bank's Supply Function and the Equilibrium Quantity of Money." Canadian Journal of Economics 43 (1970):357-385.

Pomareda, C. "Investigaciones Sobre el Seguro Agrocrediticio e Implicaciones para su Expansiôn en Amêrica Latina." Desarrollo Rural en Las Amêricas, 13 (Sept., Dec. 1981.a):161-172.

_____. "Portfolio Composition and Financial Performance of Agricultural Insurers." AGROCRED 11.81 (1981.b) IICA, San Josê, Costa Rica.

_____. "Cyclical Behavior of Interest Rates and the Adjustment of Banks' Asset Portfolio." Unpublished paper. Texas Tech University, Lubbock, Texas, 1982.a.

_____. "La Contribuciôn de la Banca de Fomento al Financiamiento del Desarrollo Agropecuario." Proyecto de Seguro Agrocrediticio, San Josê, Costa Rica, 1982.b.

_____ and R. L. Simmons. "A Programming Model with Risk to Evaluate Mexican Rural Wage Policy." Operational Research Quarterly, 28 (1977):997-1011.

_____ and T. Fuentes. "El Efecto del Seguro Agrocrediticio sobre la Producciôn y Financiamiento del Tomate Industrial". ISA/IICA, Panama, August 1981.

Porter, R. C. "A Model of Bank Portfolio Selection", Yale Economic Essays, 1 (Fall 1961):323-359.

Ray, P. K. "A Manual on Crop Insurance for Developing Countries." FAO, Rome, 1974.

Robison, L. J. Portfolio Adjustment Under Uncertainty: An Application to Agricultural Financing by Commercial Banks. PhD Dissertation, Texas A and M University, 1975.

_____ and P. B. Barry. "Portfolio Adjustment: An Application to Rural Banking." American Journal of Agricultural Economics, 59 (May 1977):311-320.

Roumasset, J. "The Case Against Crop Insurance in Developing Countries." Philipine Review of Business and Economics, March 1978.

Savage, L. J. The Foundation of Statistics. New York, Wiley, 1954.

Saving, T. "A Theory of Money Supply with Competitive Banking." Journal of Monetary Economics, 3 (1977):289-303.

Scobie, G. M. and D. L. Franklin. "The Impact of Supervised Credit Programmes on Technological Change in Developing Agriculture." Australian Journal of Agricultural Economics, 21 (1977):1-12.

Sealey, G. A. and J. T. Lindley. "Inputs, Outputs, and a Theory of Production and Cost at Depository Financial Institutions." Journal of Finance, 32 (1977):1.251-1.266.

Simmons, R. L. and C. Pomareda, "Equilibrium Quantity and Timing of Mexican Vegetable Exports,"American Journal of Agricultural Economics, 57 (August 1975):472-479.

Subotnik, A., and J. P. Houck. "Welfare Implications of Stabilization Consumption and Production." American Journal of Agricultural Economics, 58 (1976):13-20.

Taggart, R. A. and St. I. Greenbaum. "Bank Capital and Public Regulation." Journal of Money Credit and Banking, 10 (1978):158-169.

Tewari, S. K. and J. S. Sharma. "Impact of Credit and Crop Insurance as Liquidity Management Strategies Upon Adoption of Modern Technology and Income Levels on Small Farms in India." Journal of Agricultural Economics and Development, 9 (July 1978):194-204.

Thomas, W. L. et.al. "Separable Programming for Considering Risk in Farm Planning." American Journal of Agricultural Economics, 54 (May 1972):260-266.

Tobin, J. "Liquidity Preference as Behavior Toward Risk." Review of Economic Studies, 24 (1958): 65-86.

Towey, R. E. "Money Creation and the Theory of the Banking Firm." Journal of Finance, 29 (1974): 57-72.

Van Horne, J. C. Financial Market Rates and Flows. Englewood Clifs, M. J. Prentice Hall, 1978.

Vogel, R. C.   "Rural   Financial   Market   Performance:
    Implications of low Delinquency Rates."   American
    Journal of Agricultural Economics, 63 (February
    1981):58-65.
Von Pischke,   J. D.   "Rural   Credit   Project   Design,
    Implementation and Loan Collection Performance."
    Savings and Development, (FINAFRICA), Quarterly
    Review No. 2, 1980.
Von Pischke, J. D.   and D. W. Adams.   "Fungibility and
    the Design and Evaluation of Agricultural Credit
    Projects."   American   Journal   of   Agricultural
    Economics, 62 (November 1980):719-726.
_____;   P. J. Hefferman,   and D. W. Adams.
    "The Political Economy of Specialized Farm Credit
    Institutions   in   Low   Income   Countries."   World
    Bank Staff Paper No. 446, Washington D.C.,   April
    1981.
Waugh, F. V.   "Does   the Consumer   Benefit from Price
    Instability?."   Quarterly Journal   of Economics,
    58 (August 1944).
Weintgardner, H. M.   Mathematical   Programming and the
    Analysis   of   Capital   Budgeting   Problems.
    Englewood Cliffs, N. J. Prentice Hall, 1963.
WORLD BANK/EDI.   "Colloquium in Rural Finance."   World
    Bank/EDI, Washington D.C. (September 1981):1-3.

Index

Appendixes

Appendix A.
Risk and Return Characteristics of Loans, 1974–1980

	Amount Disbursed	Interest Rate	Amount Collected	Expected Duration	Actual Duration	Net Interest	Monthly Real Rate of Interest	All Maturity Rate of Interest	R	d
RIO1										
1974	334	8.14	397	12.43	23.57	63.00	.800	10.56	496	8.5
1975	330	8.00	360	7.00	19.00	60.00	.478	6.31	528	40.5
1976	445	9.00	461	7.33	17.00	18.00	.211	2.78	461	−26.5
1977	457	8.50	478	9.80	9.20	21.00	.499	6.59	478	− 9.5
1978	160	9.00	168	6.67	9.66	8.00	.517	6.82	479	− 8.5
1979	515	10.00	545	8.50	11.00	30.00	.529	6.98	480	− 7.5
1980	900	12.00	919	3.00	3.00	19.00	.704	9.29	491	3.5
x	449	9.25	475	7.28	13.20	27.00	.534	7.05	487.5	–
RIO2										
1974	2887	8.67	3012	7.56	15.00	125.00	.289	3.39	5183	−92
1975	5252	8.94	5492	7.50	11.31	241.00	.406	4.77	5252	−23
1976	3419	8.50	3793	7.50	21.00	374.00	.521	6.11	5319	44
1977	3870	8.64	4071	7.18	9.82	201.00	.529	6.21	5324	49
1978	9999	10.00	10320	12.00	10.010	321.00	.321	3.77	5202	−73
1979	3413	10.00	3548	7.00	7.57	134.00	.518	6.08	5318	43
1980	6311	11.50	6561	4.50	7.50	250.00	.528	6.20	5324	49
x	5013	9.61	5257	7.60	11.74	235.00	.444	5.22	5275	–
RIO3										
1974	42944	8.00	44182	8.00	11.00	1239.00	.262	2.88	22261	−336
1975	15746	9.33	16563	7.00	7.67	818.00	.677	5.19	22761	164
1976	15458	9.00	15857	11.00	8.00	399.00	.323	2.58	22196	−401
1977	31903	9.50	33126	13.00	11.75	1223.00	.326	3.83	22467	−130
1978	14179	10.00	14841	10.00	26.00	662.00	.179	4.65	22644	47

1979	14000	10.00	14950	6.00	7.00	950.00	.969	6.78	23105	508
1980	17235	12.00	18120	9.00	13.00	885.00	.395	5.13	22748	151
x	21638	9.69	22520	8.43	12.06	882.00	.447	4.43	22597	-

RI11

1974	-	-	-	-	-	-	-	-	0	0
1975	-	-	-	-	-	-	-	-	0	0
1976	-	-	-	-	-	-	-	-	0	0
1977	-	-	-	-	-	-	-	-	0	
1978	591	9.00	616	5.50	5.50	25.50	.862	3.45	609	3
1979	796	10.00	608	6.00	4.00	12.00	.503	2.01	601	5
1980	581	11.00	595	6.00	3.00	14.00	.803	3.21	608	2
x	589	10.00	606	5.83	4.00	17.00	.723	2.89	606	-

RI12

1974	-	-	-							
1975	-	-	-							
1976	-	-	-							
1977										
1978	4246	9.50	4334	7.50	7.50	83.90	.276	1.98	4815	-
1979	3856	10.58	4020	8.21	8.16	164.90	.521	3.24	4898	72
1980	6064	11.00	6305	7.35	5.92	240.00	.667	4.79	4948	12
601										
x	4722	10.36	4886	7.69	7.19	164.00	.488	3.60	4887	-

RI13

1974	-									
1975	-									
1976	-									
1977	-									-
1978	16882	11.00	17685	8.50	9.00	803.100	.528	4.39	23253	70
1979	20031	11.00	20941	7.33	9.33	910.00	.487	4.06	23179	144
1980	29913	12.80	31270	8.20	6.67	1357.00	.680	5.66	23536	214
x	22275	11.60	23299	8.01	8.33	1023.00	.565	4.70	23323	-

Appendix A (continued)

	Amount Disbursed	Interest Rate	Amount Collected	Expected Duration	Actual Duration	Net Interest	Monthly Real Rate of Interest	All Maturity Rate of Interest	R		d
					CO01						
1974	417	8.00	464	6.64	19.21	47.00	.587	7.52	473	—	3
1975	425	8.41	456	10.65	16.18	31.00	.451	5.78	465	—	10
1976	401	8.44	420	8.37	8.19	19.00	.578	7.40	472	—	4
1977	531	8.89	577	8.20	18.27	46.00	.474	6.07	468		8
1978	460	8.87	493	8.81	10.50	33.00	.683	8.75	478		3
1979	287	9.21	310	8.14	8.71	23.00	.920	11.78	492		16
1980	559	10.36	594	9.18	8.64	35.00	.725	9.29	481		5
x	440	8.88	473	8.57	12.81	33.00	.631	8.08	476		—
					CO02						
1974	1254	8.00	1292	5.00	6.00	38.00	.505	6.82	1459	—	10
1975	2443	8.50	2820	14.00	37.00	377.00	.417	5.63	1443	—	26
1976	1000	8.50	1047	10.00	16.00	47.00	.294	3.97	1420	—	49
1977	1600	9.00	1750	10.00	14.00	150.00	.669	9.03	1489		20
1978	1000	9.00	1025	5.00	3.00	25.00	.8133	11.24	1519		50
1979	1000	9.25	1061	8.00	13.50	61.00	.452	6.10	1449	—	20
1980	1267	9.67	1314	6.67	5.00	48.00	.758	10.23	1506		37
x	1366	8.84	1473	8.38	13.50	106.00	.561	7.57	1469		—
					CO11						
1974											0
1975											0
1976											0
1977											0
1978	729	8.05	755	6.50	6.50	26.00	.549	4.71	783	—	14
1979	678	9.00	743	6.00	9.25	64.00	1.020	8.75	813		15

Year										
1980	836	11.00	898	7.00	10.00	62.00	.742	6.37	796	—	1
x	748	9.5	799	6.50	8.58	51.00	.770	6.61	797	—	—

CO12

Year										
1974										
1975										
1976										
1977	1619	10.00	1685	4.00	4.00	66.00	1.019	8.45	2479	—	56
1978	4700	9.50	4875	8.50	14.50	174.00	.255	2.11	2334	—	89
1979	1140	10.00	1228	8.00	8.00	88.00	.965	7.99	2465	—	42
1980	1684	10.33	1760	6.67	6.67	76.00	.677	5.61	2414	—	9
x	2286	9.96	2387	6.79	8.29	101.00	.729	6.04	2423	—	—

IT01

Year										
1974	257	8.00	276	4.20	17.70	19.00	.418	5.92	459	—	7
1975	298	8.00	323	3.33	13.33	25.00	.629	8.91	471	—	5
1976	345	8.00	374	4.83	22.67	30.00	.383	5.43	456	—	10
1977	800	8.00	937	6.00	29.00	137.00	.590	8.36	469	—	3
1978	185	9.00	191	4.00	6.00	6.00	.540	7.65	466	—	0
1979	422	9.50	435	4.38	4.50	13.00	.684	9.69	475	—	9
1980	622	10.33	641	4.67	6.00	20.00	.536	7.59	466	—	0
x	433	8.69	457	4.55	14.17	36.00	.470	7.65	466	—	—

IT02

Year										
1974	1000	8.00	1212	10.00	53.00	212.00	.4	5.12	1702	—	40
1975	2000	8.00	2065	4.00	6.00	65.00	.542	6.94	1731	—	10
1976	1800	8.00	1870	6.00	7.00	70.00	.555	7.10	1734	—	7
1977	2000	8.00	2060	6.00	6.00	60.00	.5	6.40	1723	—	18
1978	2000	10.00	2081	6.00	6.00	81.00	.675	8.64	1759	—	18
1979	1000	10.00	1031	6.00	6.00	31.00	.517	6.62	1726	—	15
1980	1531	11.00	1611	3.40	5.60	81.00	.945	12.10	1815	—	73
x	1619	9.00	1704	5.91	12.80	86.00	.590	7.56	1741	—	—

Appendix A (continued)

	Amount Disbursed	Interest Rate	Amount Collected	Expected Duration	Actual Duration	Net Interest	Monthly Real Rate of Interest	All Maturity Rate of Interest	R	d
					ITl1					
1974										
1975										
1976										
1977	–									
1978	150	9.00	157	6.00	6.00	7.00	.777	3.50	408	1
1979	433	9.00	448	4.67	4.00	15.00	.866	3.90	409	2
1980	600	9.50	612	4.50	3.50	12.00	.571	2.60	404	– 3
x	394	9.17	403	5.06	4.50	11.80	.738	3.33	407	–
					ITl2					
1974										
1975										
1976										
1977	–									
978	1500	12.00	1517	5.00	5.00	71.00	.947	4.57	1670	9
1979	1000	10.00	1044	5.00	5.00	44.00	.88	4.25	1665	4
1980	2290	10.50	2358	7.00	4.50	88.00	.659	3.18	1648	– 13
x	1597	10.83	1658	5.67	4.83	61.00	.829	4.00	1661	–
					ITl2					
1974	502	8.09	535	5.27	12.73	33.00	.516	6.04	536	3
1975	747	8.25	772	6.00	6.50	25.00	.515	6.02	535	2
1976	619	8.67	655	11.00	26.67	36.00	.218	2.55	518	– 15
1977	290	8.29	313	5.14	12.57	23.00	.631	7.38	542	9
1978	437	8.89	455	6.11	7.89	17.00	.493	5.77	534	1
1979	481	9.00	495	4.43	9.71	14.00	.299	3.49	523	– 10

1980	463	10.00	481	5.62	5.87	18.00	.662	7.74	544	10
x	505	8.75	530	6.22	11.70	24.00	.476	5.57	533	-
					VG02					
1974	2759	8.33	2881	6.33	7.67	121.00	.572	5.70	2481	- 3
1975	2391	8.91	2569	7.54	12.73	178.00	.585	5.83	2484	0
1976	1644	9.00	1731	6.00	10.56	87.00	.501	4.99	2464	- 20
1977	2275	9.50	2388	6.00	9.25	113.00	.587	5.35	2472	- 12
1978	1843	9.00	2033	6.00	17.00	190.00	.606	6.08	2520	36
1979	3394	10.20	3441	5.00	6.40	47.00	.216	2.15	2397	- 87
1980	2325	10.67	2460	5.22	6.11	135.00	.950	9.46	2569	85
x	2376	9.37	2500	6.01	9.96	158.00	.565	5.65	2484	-
					CF01					
1974	489	8.00	522	9.47	14.74	33.00	.458	5.76	518	- 5
1975	481	8.00	514	10.12	11.81	23.00	.405	5.09	515	- 8
1976	566	8.00	612	13.50	19.75	47.00	.420	5.28	516	- 7
1977	412	8.14	436	11.57	10.86	24.00	.536	6.74	523	0
1978	553	8.71	599	11.14	12.00	46.00	.693	8.72	533	10
1979	300	9.09	317	10.91	9.73	17.00	.582	7.32	526	3
1980	630	10.20	669	9.00	9.20	39.00	.673	8.47	531	8
x	490	8.59	524	10.81	12.58	33.00	.538	6.77	523	-
					CF02					
1974	2811	8.00	3001	10.21	17.50	189.00	.384	4.80	2564	- 46
1975	2846	8.00	3062	10.00	18.33	216.00	.414	5.18	2574	- 36
1976	1806	8.00	1907	11.33	15.00	101.00	.373	4.67	2561	- 49
1977	3224	8.50	3318	7.00	5.50	94.00	.530	6.63	2609	1
1978	2097	9.67	2254	14.00	14.00	157.00	.535	6.69	2611	1
1979	1906	10.37	2036	12.12	10.00	130.00	.682	8.53	2656	46
1980	2439	11.37	2583	9.00	7.25	144.00	.814	10.18	2696	86
x	2447	9.13	2590	10.52	12.51	7.00	.533	6.67	2610	-

Appendix A (continued)

	Amount Disbursed	Interest Rate	Amount Collected	Expected Duration	Actual Duration	Net Interest	Monthly Real Rate of Interest	All Maturity Rate of Interest	R	d
LV01										
1974	700	8.00	817	51.00	36.00	117.00	.464	13.14	691	- 20
1975	450	8.00	531	40.00	26.00	81.00	.692	19.60	731	- 20
1976	500	8.00	594	49.00	49.00	94.00	.384	10.88	677	- 34
1977	700	9.00	761	8.00	16.00	61.00	.545	15.44	705	- 6
1978	519	12.00	581	18.00	22.00	62.00	.643	15.38	705	- 6
1979	800	10.00	942	24.00	21.00	142.00	.845	23.94	755	45
1980	-	-	-	-	-	-	-	-	-	-
x	611	9.17	704	31.33	28.33	93.00	.579	16.40	711	-
LV02										
1974	3029	8.48	3878	60.05	57.48	848.00	.487	16.19	4637	- 96
1975	1831	8.75	2364	69.50	51.25	533.00	.568	18.88	4741	8
1976	3690	8.71	4129	49.29	42.71	439.00	.278	9.24	4336	- 337
1977	2090	9.00	2258	11.00	11.00	168.00	.749	24.89	4981	248
1978	5488	10.67	6215	21.33	22.00	727.00	.602	20.01	4786	53
1979	7800	12.00	8600	13.00	15.00	800.00	.684	22.74	4895	162
1980	-	-	-	-	-	-	-	-	-	-
x	3988	9.50	4574	37.36	33.24	586.00	.561	18.66	4733	-
LV12										
1974										
1975										
1976										
1977										
1978	4415	12.00	4887	23.00	13.00	472.00	.822	12.39	5658	28
1979	5653	12.17	6379	24.33	17.17	726.00	.748	11.28	5602	- 28

	5034	12.08	5633	23.66	15.08	599.00	.785	11.83	5630	—	
1980 / x											
T01											
1974	531	8.00	552	6.86	7.29	21.00	.542	8.26	555	—	1
1975	280	8.00	302	56.00	56.00	22.00	.140	2.13	524	—	32
1976	579	8.00	599	5.33	9.33	21.00	.389	5.93	543	—	13
1977	534	8.80	568	6.60	9.60	34.00	.663	10.10	565		9
1978	540	9.22	581	13.22	9.89	41.00	.768	11.70	573		17
1979	567	10.00	591	7.80	6.80	25.00	.648	9.87	564		8
1980	560	10.80	589	13.20	7.80	30.00	.687	10.47	567	—	11
x	513	8.97	540	15.57	15.24	28.00	.548	8.35	556		—
T02											
1974	1886	8.00	2044	23.60	19.80	158.00	.423	5.66	2595	—	42
1975	1999	8.80	2170	6.80	21.60	131.00	.396	5.30	2586	—	41
1976	2768	9.00	2850	5.00	6.00	82.00	.494	6.61	2618	—	19
1977	2679	9.00	2954	11.00	13.00	275.00	.789	10.56	2715		78
1978	2919	10.00	3263	18.12	16.60	343.00	.712	9.53	2690		53
1979	1677	10.00	1743	7.00	8.33	66.00	.472	6.32	2611	—	26
1980	3265	10.00	3422	8.00	8.50	157.00	.566	7.58	2642		5
x	2456	9.261	2635	11.36	13.39	179.00	.550	7.36	2637		—
T11											
1974											
1975											
1976											
1977											
1978	865	10.00	902	5.00	9.00	37.00	.475	3.17	715	—	3
1979	648	10.00	671	5.00	6.00	22.00	.566	3.77	719		1
1980	565	8.00	582	5.25	5.00	12.00	.602	4.01	721		2
x	693	9.67	718	5.08	6.67	25.00	.548	3.65	718		

Appendix A (continued)

	Amount Disbursed	Interest Rate	Amount Collected	Expected Duration	Actual Duration	Net Interest	Monthly Real Rate of Interest	All Maturity Rate of Interest	R	d
					TL2					
1974										
1975										
1976										
1977										
1978	4987	10.00	5197	5.00	8.33	210.00	.505	4.47	3868	11
1979	3210	10.43	3435	5.29	13.00	225.00	.539	4.47	3879	-
1980	2913	9.78	3001	5.44	5.22	88.00	.579	5.12	3892	11
x	3703	10.07	3878	5.24	8.85	174.00	.541	4.79	3879	-
					RA03					
1974										
1975	28538	8.00	29421	11.50	15.50	882.00	.199	2.74	37249	- 671
1976	56438	8.00	57500	30.00	8.00	1062.00	.235	3.24	37431	- 490
1977										
1978	41561	8.20	43435	8.00	22.80	1874.00	.198	2.73	37246	- 675
1979	44397	8.00	48322	13.00	16.67	3925.00	.530	7.31	38906	985
1980	10344	10.00	10656	8.00	6.00	312.00	.503	6.94	38772	851
x	36256	8.44	37867	14.0	13.79	1611.00	.333	4.59	37921	-

Appendix B.
Coding for the Linear Programming Model, BANK

The coding used to define rows and columns in the model uses eight characters to define each vector. In the rows section the name is preceded by an indication of the inequality as follows:

> N   indicates no restriction
> G   indicates greater or equal ($\geq$)
> L   indicates less or equal ($\leq$)
> E   indicates equal (=)

The eight characters are interpreted as in the following example for the fourth row:

> 1   A   amortization
> 2-3   BF   borrowed funds
> 4   .   blank space
> 5   .   blank space
> 6   .   blank space
> 7-8   01   year of the model

In similar form in the case of the first column for example:

> 1   T   transfer activity
> 2-3   LR   loan recovery
> 4-5   00   period of origin
> 6   .   always blank
> 7-8   .   year of the model

The coding is the following:
Block 1.

Type of activity or row

> L   loan
> B   borrowing
> A   amortization
> C   cost
> I   investment
> T   transfer activity
> R   restriction
> E   balance equation
> D   deposit

Block 2.3

Name of activity and institution

Loans for:
RI   rice
CO   corn - sorghum

Appendix B (continued)

Loans for:
IT industrial tomatoes
VG vegetables
CF coffee
LV livestock
OT other

Investments
BD bonds

Institutions
ID Interamerican Development Bank
WB World Bank
AI AID
CB commercial banks

Financial variables

LR loan recovery
LE leverage constraint
IE interest earnings (on loans)
GS government subsidy
OR other resources
OE operating expenses
OL outstanding loans
CD capital disbursements
SA savings
CH checking
IP number of insured loans
IC value of insured crops
IL value of insured livestock
CP crops loan contraints
LV livestock loan constraints
IN interest payments on borrowed funds
NR net returns
PN resources office staff requirements
VH vehicle requirements
LT loan officer time
CT collection officer time

Block 4

Use of insurance

0 = No
1 = Yes

Appendix B (continued)

Block 5

Type of loan, bonds, saving and checking

Loans
1 = small (⩽ $1,000)
2 = medium ($1,000 - $10,000)
3 = large (⩾ $10,000)

Bonds
1 maturing in one year
2 maturing in two years

Checking
1 of average balance equal to $10,000
2 of average balance equal to $100,00

Savings
1 of average balance equal to $1,000
2 of average balance equal to $10,000
3 of average balance equal to $100,000

Blocks 4.5

Year of origin of funds

01, 02,...10

Block 6

Blank (always)

Blocks 7.8

Year of destination or year of collection of funds

01, 02,...10